AVID

READER

PRESS

PIPE
Dreams

The Urgent Global Quest to
TRANSFORM THE TOILET

CHELSEA WALD

AVID READER PRESS
New York   London   Toronto   Sydney   New Delhi

AVID READER PRESS
An Imprint of Simon & Schuster, Inc.
1230 Avenue of the Americas
New York, NY 10020

First Avid Reader Press hardcover edition April 2021

AVID READER PRESS and colophon are
trademarks of Simon & Schuster, Inc.

For information about special discounts for bulk purchases,
please contact Simon & Schuster Special Sales at 1-866-506-1949
or business@simonandschuster.com.

The Simon & Schuster Speakers Bureau can bring authors to
your live event. For more information or to book an event contact the
Simon & Schuster Speakers Bureau at 1-866-248-3049
or visit our website at www.simonspeakers.com.

Interior design by Kyle Kabel

Manufactured in the United States of America

1   3   5   7   9   10   8   6   4   2

Library of Congress Cataloging-in-Publication Data has been applied for.

ISBN 978-1-9821-1621-7
ISBN 978-1-9821-1623-1 (ebook)

For Cyril and Ephraim

Contents

A Toolik tower in 2010.

Preface

At Toolik Field Station in Alaska, it's easy to find the toilets: look up. In the flat, treeless arctic tundra, they're the highest thing around. They're basically outhouses, but instead of sitting on the ground, they're perched on holes over thousand-gallon collection tanks. Walking up the outdoor wooden staircase is as good as announcing to a hundred fellow campmates, "I've gotta go."

It was while sitting on one of these holes in "the tower" during a journalism fellowship at the scientific research camp in 2010 that I first began to contemplate the quantity of pee and poop that a person can create—it was right there underneath me. I later found out how much: according to a synthesis of findings from around the world, about 100 pounds of poop and about 140 gallons of pee per person each year. At Toolik, where peak summer capacity is about 160 people, all that crap is a big problem. Not only does it begin to stink, but also the scientists can't risk contaminating the pristine landscape—which is what they are there to research, after all—with their own waste. So every few days, a vacuum truck drives in on the Dalton Highway, sucks out the tanks, and hauls the contents far away.

Later, when I started to write articles for magazines about innovations in toilets for people who don't have safe sanitation, I thought of Toolik's poop problem in a new light. In many ways, it was just a microcosm of the world's poop problem. Modern sanitation infrastructure has created the illusion that our excreta just disappear like magic, a phenomenon dubbed "flush and forget." But poop doesn't just drive down the highway into the sunset. Although it breaks down, its components live on and must go somewhere. Too often that's to the wrong places, where they can poison people or ecosystems—whether in distant slums or in America's cities, small towns, and wild areas.

But I also learned that people have been devising clever solutions for the poop in our midst since the dawn of time and we are again recasting toilets as an opportunity. I have become captivated by this vision and decided to write a book about it, because it's not just about getting toilets to everybody—though that must happen, and quickly—but also about shepherding a new generation of them into the world that allows our bodies, our planet, and our societies to be far healthier.

When I started drafting this book in earnest in 2018, the realist in me found this vision sometimes exceedingly optimistic—hence the title *Pipe Dreams*—though worth sharing in the hopes that a wider public would include toilets in their own fights for justice, health, and restoring planetary balance. But as I finish it in 2020, when the coronavirus pandemic is overtaking our tragically unhealthy world, the vision seems far more urgent and somehow also far more achievable. The home we have built for ourselves on this planet is falling apart in so many ways. Why not start the renovation with the toilet?

Let's get something out of the way. I know it's weird to read a book all about toilets. It's still sometimes strange to me that I'm *writing* a book about toilets. As a child, I was not a fan of potty humor (for which I've now developed a fondness, if possibly not a knack), and I fortunately never had any medical condition that kept me in the bathroom for an extended period of time. Years ago, before I began this work, a delightfully weird friend tried to strike up a conversation about why toilets in Germany have a shelf in them where poop lands before flushing. Why would we talk about *that*? I remember thinking, shooting her an appalled look. To toilets, the lid on my mind was closed.

That is where most people start. Many of today's toilet revolutionaries, regardless of their field, can tell a story about the moment that they opened their minds to toilets. Engineer Emily Woods started out working on water treatment systems for the world's poor but soon came to understand that the contamination in the water she was treating originated with the lack of safe toilets, so she went on to co-found a sanitation business in Kenya. ("We always joke that people join the poop world for fame and prestige," she has said.) Raul Pacheco-Vega, a Mexican-Canadian political scientist and geographer, was doing fieldwork in a small town in Mexico when he felt the call of nature. "I spent literally an hour walking around town knocking on doors because I could not find a toilet where I could feel comfortable," he says. That's when he concluded that having a toilet wasn't enough. "You may have access to a toilet, but you may not want to use it."

Mwila Lwando, a Zambian toilet entrepreneur, was visiting the country's food markets with the idea of creating new information products for farmers when he noticed that those markets had no place for workers to relieve themselves. "Whenever we're here and we're trying to use the toilet, we can't find one. And it means that's not only us; it means there's sixty thousand other people who need to use it." So he changed course, instead providing clean, attractive pay toilets that recycle water through the system and generate electricity for lighting from biogas. He predicted a "sanitation gold rush"—and he wanted to get in on it.

Toilet archaeologist Gemma C. M. Jansen told me that it took her two decades before she figured out that she needed to imagine real people actually using the ancient Roman toilets she was studying. What is the path that they walked? What could they see? How comfortable was the seat? How would they have held their clothes? She had been inadvertently granting the ancients the same privacy that she would her friends. Only when she changed her perspective did she realize that some people wrote toilet graffiti from a sitting position, while others from a standing one—a sign, perhaps, that, people used the toilets in both ways.

For me, the process of opening my mind happened in stages. After Toolik, I started to see the stories where toilets were the main subject, often about places with little or failing sanitation infrastructure. But then I started to see that, because everybody poops, toilets play a role in every big story in the world. I followed news—some prominent, some buried—about overwhelmed wastewater treatment plants after hurricanes, poop piling up in national parks during a major government shutdown, sewage as a source of microplastic pollution, the toilet paper shortage that could occur after Brexit, and even the tale of the private toilet that accompanied North Korea's Chairman Kim Jong-un to a meeting with President Donald Trump.

Finally, I came to realize that, until we consider toilets, we can't understand any story, including our own. In the blockbuster film *Hidden Figures*, based on real NASA mathematicians in the 1960s, the brilliant protagonist Katherine Johnson, who was Black, has to run-walk her way across the campus, in pumps, to get to a "colored" women's restroom. This routine trip puts her job at risk since her clueless white male bosses wonder why she is so rarely at her desk. Although the specific anecdote may be more fiction than fact, it gave me a new understanding of the practical realities of life under Jim Crow. Flies on the walls for millennia, toilets can be a veritable window to a greater understanding of almost every aspect of history, culture, and society, if only we take a look at them.

When I reflect on my own life, one particular toilet comes to mind. It illustrates the complex, layered meanings we assign to these places of elimination, at the same time we try to put them out of mind. As an undergraduate, I spent a lot of time in Columbia University's physics building, Pupin Hall (which, I should add, is *not* pronounced "poopin' hall"). Like many of its era, the historic brick structure had only one restroom per floor. Most floors had men's rooms; only on a few floors were there women's. The one I used was practically luxurious, unlike any other on the whole campus, equipped with a chaise longue and generous piles of free sanitary products. Yet, as I remember, hardly anyone was ever in there. That was lovely in a way, and I sometimes used it to rest and think and just get a break. At the same time, however, every trek to that frilly ghost toilet managed to remind me that my uterus made me a rarity, maybe even an oddity, in the world of physics—a world, in the end, I chose not to enter.

About a decade later, those same restrooms became a flash point in a different but related cultural transformation. By then, they had gotten reapportioned in a more welcoming (and, one hopes, representative) way: six for men, five for women, and a gender-neutral bathroom. But, in 2017, news spread that a physics Ph.D. student had torn down LGBTQ-friendly signs, including one that read: "Please use the restroom that is most consistent with your gender identity." The student who reported it saw the vandal "give a look of disgust, and then rip it off and throw it away," he told the *Columbia Spectator*. This is just one of the ways in which, as Slovenian philosopher Slavoj Žižck put it, "as soon as you flush the toilet, you're right in the middle of ideology."

The lid on your mind may be already pretty open to the topic of toilets, or it may be just starting to open. I've come a long way since the time I refused to engage with my friend. Now I'm the one starting uncomfortable conversations and telling awkward stories. (Later, I'll even plunge into why German toilets have those poop shelves.) Because, as I've learned, there's nothing to fear when we talk about toilets. There's only something to fear when we don't.

A note on terms. A lot has been written about the history of the toilet's many euphemisms. "The toilet is a physical barrier that takes care of the physical dangers of excrement. Language takes care of the social ones," writes British journalist Rose George in her classic 2008 book on sanitation, *The Big Necessity*:

> Water closet. Bathroom. Restroom. Lavatory. Sometimes, we add more barriers by borrowing from other people's languages. The English took the French *toilette* (a cloth), and used it first to describe

a cover for a dressing table, then a dressing room, then the articles used in the dressing room, and finally, but only in the nineteenth century, a place where washing and dressing was done, and then neither washing nor dressing. (They also borrowed *gardez l'eau*, commonly shouted before throwing the contents of chamber pots into the streets, and turned it into "loo.") The French, in return, began by calling their places of defecation "English places" (*lieux à l'anglaise*) and then took the English acronym *WC* (water closet) instead. The Japanese have dozens of native words for a place of defecation but prefer the Japanese-English *toiretto*. You have to go back to the Middle Ages to find places of defecation given more accurate and poetic names: Many a monk used a "necessary house." Henry VIII installed a House of Easement at his Hampton Court Palace. The easiest modern shorthand for the disposal of human excreta—sanitation—is a euphemism for defecation which is a euphemism for excretion which is a euphemism for shitting.

Indeed, *sanitation* as a term is also difficult, since you may know it as the general collection of trash, as in New York City's Department of Sanitation. But I won't use that meaning in this book, and I'll prefer the straightforward *toilet*.

As for the foul stuff itself, *shit* has become a term of art among experts, be they archaeologists, engineers, or sociologists, and I'll use it a little in this book. "It is a word with noble roots, coming from a family of words that also contains the Greek *skihzein*, the Latin *scindere*, or the Old English *scitan*, all meaning, sooner or later, to divide or separate," George writes. Today, however, it carries loads of meanings beyond the literal: as a noun, it can refer to objects ("Is that your shit?"), people ("You're a little shit"), or experiences ("That's some crazy shit"), and it's also multipurpose as a verb ("I do not shit you"), as an adjective ("What shitty weather"), and as part of a wide variety of phrases ("shit a brick," "tough shit," "full of shit," and so on).

Excrement, *feces*, *ordure*, and the like seem too formal, even jargon-y, and inappropriately sterile on top of that. Then there are the euphemisms, but *human waste* denies the value inherent in it, *manure* sounds like something from cows, while *humanure* hasn't caught on because, well, it's silly. "There is no neutral word for what humans produce at least once a day, usually unfailingly. There is no defecatory equivalent of the inoffensive, neutral 'sex,'" George writes.

There's one euphemism that George doesn't evaluate, however, and doesn't use except when quoting others: *poo*, which in American English tends to get rendered as *poop*. About three centuries ago, it morphed from an onomatopoeic meaning of "to make a short blast on a horn" to "to break wind softly"—that is, to fart. From there it made the relatively easy transition to a children's euphemism for excrement and the act of passing excrement. It is the case that a decade ago the term sounded nearly as silly as *humanure* and also nearly as taboo as *shit*. But Google Trends shows that, in the United States, interest in the term *poop* has steadily grown over the years, the word becoming almost four times as popular as it was in 2008, suggesting that society's relationship with the term has shifted. (*Shit*, on the other hand, has held steady.)

A short history of the "pile of poo" emoji—which looks, on first glance, like a smiling swirl of chocolate soft-serve ice cream—is instructive. The symbol started in Japan in the 1980s as a reference to a joke from a popular manga series; in the late 1990s, it became one of the original emoji. (In Japan, "golden poo" charms are also a symbol of good luck because of a similarity in the sound.) In 2008, the pile of poo left Japan as part of the first Gmail emoji set, as a brown turd with yellow odor lines and flies flitting around it, despite executives' worries about offending their customers outside Japan. Over time, though, the poo emoji got increasingly bigger eyes and a friendlier grin—ultimately it was so cute that people not only used it in their messages but also bought plush toys of it. People had started to like poop! For this reason—and also because *poop* goes with *pee* while *shit* goes with the unpoetic *piss*—*poop* and *poo* have become my preferred terms.

In part due to the abundance of euphemisms, people who work in this field tend to acquire nicknames, whether they want them or not. In the old days, there was the Groom of the Stool, the trusted and powerful male courtier responsible for assisting the king with his toileting, possibly including wiping the monarch's bottom. Today, there's the Poo Princess, the Queen of Latrines, Potty Girl, Mr. Toilet, Dr. Fatberg, Sewer Chic, Loo Lady, and the Puru. I have some nicknames, too. One is the Queen of Loo-topia—for reasons you'll see in the coming chapter. Another was given to me by Dutch friends in The Hague, where I live. Here, as the story goes, before indoor plumbing posh people had fancy commodes. Still today the city's preppies—those who pop their collars, drape their sweaters around their shoulders just so, pronounce their *r*'s at the front of their mouths instead of the back, and play field hockey at private schools with members of the royal family—are known as *Haagse kakkers*, or Hague shitters. I'm not known to hobnob with the Dutch elite, but, due to my odd interest in toilets, my friends thought I should be called a *Haagse kakker*, too.

I know that not everybody is as at ease with this topic as I am by now, after years of immersing myself in poop (fortunately not literally). Sometimes I forget just how uncomfortable it can make people. If you experience any discomfort while reading, please don't let it stop you, because there is so much to gain by joining this conversation. We need to be willing to overcome long-standing taboos against thinking and talking about toilets, rather than flushing these uncomfortable thoughts away.

People joke about having eureka moments while on the toilet. But, looking back, I guess it really did happen to me. My multiple daily climbs up the towers at Toolik launched me on one of the most fascinating journeys of my life. Please join me up the wooden staircase to reconsider the toilet.

The New Toilet Revolution

Are you ready?

> To keepe your houses sweete, clense privie vaultes.
> To keepe your soules as sweete, mend privie faults.

> —Sir John Harington,
> *A New Discourse of a Stale Subject,*
> *Called the Metamorphosis of Ajax* (1596)

Getting to Loo-topia

I've come to the north of the Netherlands, to a medieval town called Sneek, population thirty-three thousand, whose name is pronounced with a hard *a*, like the tubular animal that tempted Eve with an apple, and not with a hard *e*, like the act of slinking around. An imposing seventeenth-century city gate, called the Watergate, sits astride one of the town's many canals, and, in the summer, major regattas take over a nearby large lake, called the Sneekermeer.

But Sneek has a lesser-known tourist route, one for people like me, who are interested in the unsavory side of water. In 2004, a company called DeSaH (the Dutch acronym for Decentralized Treatment and Reuse) installed vacuum toilets, not unlike those you might have used

1

on airplanes or trains, in a thirty-two-house complex. It treated the waste from those toilets in one garage instead of sending it through the sewers to a centralized treatment plant. The successor project, called Waterschoon (*schoon* means "clean" in Dutch), serves more than two hundred apartments. The now-king Willem-Alexander of the Netherlands himself came to open it in 2011 when he was still known as the Water Prince for his advocacy regarding all things Dutch and watery.

These "New Sanitation" projects are just a tiny corner of a new, global toilet revolution—a growing movement that has the goal of upending the way we manage our most basic bodily functions. The toilet we have today was born in England during the nineteenth-century Victorian sanitary revolution. It was a modern miracle: the ubiquitous user interface of a vast, reliable system of pipes, sewers, and, later, treatment plants that was both extraordinarily convenient and medically essential, stopping outbreaks of devastating diseases such as cholera and typhoid fever and giving countless people longer, more productive lives. In 2007, the readers of the *British Medical Journal* named the sanitary revolution "the most important medical milestone since 1840," putting it ahead of antibiotics, anesthesia, and vaccines. In 2013, the *Economist* featured a flush toilet on its cover, with Rodin's *The Thinker* sitting on it, wondering, "Will we ever invent anything this useful again?"

But our toilets no longer look quite so miraculous as they once did. While our societies have changed dramatically over the past century, we have allowed our toilet systems to stagnate. Today, they are inadequate to the challenges ahead of them. Our cities are growing, overwhelming their aging, inflexible infrastructures, especially during shocks such as storms. We're throwing more and more junk into the system, from trash to oil to pharmaceuticals to toxic chemicals. And resources are growing ever scarcer, making the squandering of the water, nutrients, energy, and other components in wastewater harder to stomach. American inventor and visionary R. Buckminster Fuller could have been talking about the toilet when he wrote in 1970 that "pollution is nothing but the resources we are not harvesting. We allow them to disperse because we've been ignorant of their value."

On top of that, despite some progress toilets fail to reach so many. According to the latest numbers from 2017, some 2 billion people still lack a minimally adequate toilet and hundreds of millions don't use a toilet at all. Most of these are in low- and middle-income countries, where toilet systems fail to serve both rural and urban areas. Many cities are growing much faster than infrastructure can keep up with, and they can get so crammed in parts that there's no room for private toilets. Sometimes sewage flows through open trenches instead of closed pipes. And people may resort to using "flying toilets"—poop-filled plastic bags that they discard in streets or ditches, or just fling away.

You might be surprised by other places that don't have decent toilets. According to one report on the state of water and sanitation in the United States, "It is safe to say that more than two million Americans live without complete plumbing" and "even that may be an underestimation." Accurate numbers aren't available in part because, in 2016, the U.S. Census Bureau's American Community Survey eliminated its question about toilet access. Lack of safe sanitation, as environmental health activist Catherine Coleman Flowers puts it, is "America's dirty secret."

Some people in the Black Belt region of the South, where Flowers hails from, unable to buy or maintain septic tanks specialized for the clay soil, "straight pipe" their wastewater just a few feet from their homes, into yards where children sometimes play. Researchers found that two in five study participants from one of these poor areas in Alabama had intestinal parasites, particularly hookworm, which today are mostly associated with low-income countries. In some rural Alaskan communities, where the ground isn't suitable for wells or septic systems, thousands of people defecate into euphemistically named "honey buckets," which they then empty by hand into very unsexy "lagoons." In those communities, gastrointestinal, respiratory, and skin infections are rampant. In Hawaii, eighty-eight thousand mostly unlined cesspools—little more than holes in the ground—accept wastewater from toilets and other household plumbing, which then leaches out and contaminates local streams, groundwater, and the ocean with pathogens. And just a few years ago, San Diego's homeless community, then the fourth largest in the United States, fell victim to a hepatitis A outbreak, which sickened 592 people and killed 20, in part due to conditions such as insufficient toilets and handwashing facilities in the encampments.

Around the world, the toilet has become a paradox. Lauded as a savior of civilization, it also exacerbates many of the world's problems: inequality, disease, pollution, climate change, water shortages, soil degradation, waste. It's time, many think, to escape the old paradigms and harness toilet systems not only to sequester poop but also to do a variety of other desirable tasks, such as guarantee everyone a place to go, watch for disease outbreaks, make fertilizer and fuel, manufacture components for bioplastics and asphalt, synthesize drugs, produce clean water, and build job-creating businesses. Bill Gates, the computer pioneer turned mega-philanthropist and toilet enthusiast, delivered a widely seen speech about "radically new . . . alternatives for collecting, managing, and treating human waste," while he was wearing a suit and holding a beaker of shit (his own, one hopes—or doesn't?). The focus of the Bill and Melinda Gates Foundation is on the world's poor, but the revolution is needed everywhere, poor or rich, cold or hot, sewered or not.

New Sanitation in Sneek is just one attempt to step away from conventional toilet systems, but it's a good introduction to the revolution because it is a sort of photo negative of those. Instead of diluting poop and pee with water, it concentrates them; instead of handling poop and pee centrally, it does it locally; instead of primarily using resources to treat poop and pee, it makes resources out of them.

The process begins with the vacuum toilets. I try one out at the company's small headquarters in an industrial park. For research purposes, they have installed the newest generation there, but only into their men's room. I want—and fortunately also have a need—to use them, so my host, Moniek Agricola, the young engineer who takes care of much of DeSaH's day-to-day maintenance, stands guard outside. They're not like the metal toilets you'd see in an airplane but are instead fashionably white and curvy, with blue glowing buttons for flushing. There's only a tiny amount of water in the bowl, which serves mainly as a security blanket for users. I sit down and make my offering. When I press the button, a hole at the bottom opens and the void, generated by a pump, sucks everything down with a *whoosh!* that is gentler than that on an airplane, even if I leave the lid open (which I'm not supposed to do).

In this damp region, the point of the extremely low-flow fixtures isn't primarily water conservation, although a recent drought shocked the Dutch into rethinking their trust in the never-ending abundance of water. New Sanitation bases its treatment process on microbe-powered *anaerobic digestion*, which benefits from the vacuum toilets' concentrated waste, rather than the *aerobic digestion* that forms the basis of conventional wastewater treatment. As you might know from your workouts, aerobic processes demand oxygen, and anaerobic ones don't. Anaerobic digesters for sewage can save energy because they don't have to pump air to the microbes that do the work. They also take up less space and leave less residue behind.

They have other benefits, too, as I see when Agricola takes me to the Waterschoon site, in a quiet complex that's built of red bricks, solid and low to the ground, with some medieval-style arches for flair. The residents, who must qualify for government housing subsidies to live here, display knickknacks and grow plants in large windows. Pipes carry the toilet waste, known as "black water," from about two hundred apartments to a treatment building at the center of the complex. Entering, I hear the loud hiss of machinery and my nose registers the sharpness of ammonia layered over an earthy richness. In a corner, the black water flows upward through a two-story black metal cylinder containing a blanket of microbes. These microbes, like the ones that thrive in lake sediments and the stomachs of cows, need little to no oxygen to live. As they munch on the waste, they produce biogas, which flows in short puffs to a boiler that heats water for the complex, replacing some of the natural gas that the Netherlands plans to phase out in the coming years because extracting it has caused a series of small but damaging earthquakes. The little solid matter they leave in their wake empties from a hole in the bottom when it gets to be too much. In a way, the digester is like an artificial stomach, burping and pooping.

The digester also produces a liquid, which then enters two more reactors that deal with the nitrogen and phosphorus compounds in it. British anthropologist Mary Douglas famously described dirt as "matter out of place," and so it is with these nutrients: they can be either pernicious polluters or essential fertilizers, depending on where people choose to put them. If released from wastewater treatment plants indiscriminately, these nutrients can lead to overgrowth of algae in waterways, suffocating the other forms of life in there. But at Waterschoon, the first reactor uses specialized microbes to pull out nitrogen, which is bound up in the compound ammonium, turning it into a gas and releasing it into the atmosphere, which is already 78 percent nitrogen. In the second, a chemical reaction binds the phosphorus into a mineral called struvite, which emerges from the reactor as small, smooth stones that can be used as a fertilizer.

The system also treats toilets as part of a larger cycle of water and waste. The Waterschoon housing units come with in-sink food grinders (what many Americans call by the product names Disposall or InSinkErator but tend to be banned in Europe), which also feed into the anaerobic reactor, providing the microbes with additional nourishment. And the homes' shower, sink, dish- and clothes-washing water—"gray water"—flows through a different set of pipes to join the liquid from the black water process for a final set of cleaning steps. That gray water is relatively hot, so the facility also uses a heat exchanger to transfer some of the energy to the water used for heating the homes. Ultimately, DeSaH would like to clean the water further and cycle it back into the houses for toilet flushing and other uses—even possibly for drinking—but for now the system discharges it into the sewer.

Biogas, fertilizer, heat, and water savings in a modular, flexible system. Who wouldn't want toilets like these? Later, when I pose that question in a meeting with project collaborators from the company, the municipality, and the water board at DeSaH's headquarters, I learn that this concept has been a harder sell than you might think. Sneek has hosted a parade of interested delegations from around the world—as far as China—but almost never do visitors go home and install New Sanitation. Faced with financial, regulatory, and cultural headwinds, says Flip Kwant, DeSaH's commercial director, "they never came back." It was even a hard sell in the Netherlands: before Sneek, two other new housing developments in different Dutch cities considered but ultimately refused the opportunity to pilot the system. Wouldn't the vacuum toilets be too loud, too smelly, too *embarrassing*? people wondered. For a sanitation engineer, or even a designer or sociologist or businessperson or economist or philanthropist, to be part of the revolution already takes a certain fortitude, since they must cope with disgusting materials, limited funding opportunities, risk of acute and chronic disease, and terrible poop jokes. Unlike with computers and smartphones, most people don't want to be early adopters of new toilet technologies. They just want reliable toilets that work.

But Sneek was special. It wanted to bolster its reputation as a "Water City," the housing developers thought it would give the new project cachet, and DeSaH's presence in the community built trust. What's more, the locals have grit: the day I'm there, a major windstorm blows through, putting the nation (and me) on high alert, but the seafaring people in this region are used to much worse. They've just pulled their boats up onto shore and gotten on with life. So it doesn't surprise me that Waterschoon has drawn few complaints and that one neighbor in the original thirty-two-house project, when researchers questioned him about his openness to participating, just said, Why not? "They have to experiment *somewhere*."

Yes, they do—and they will, as the worldwide movement generates a motley mix of concepts for the future of toilets. The story of the toilet revolution is not just a tech story: it's a culture story; it's a people story. And, even if you don't work in toilets yourself, in the coming years the revolution may ask something of you. To pee and poop in separate holes. To compost your toilet waste and spread it in your garden. To cook with gases extracted from a septic system. To eat off paper plates made from recycled used toilet paper. To talk to your toilet about your health. To drink cleaned-up toilet water. To talk to your friends—and strangers—about your toilet. Are you ready? Are you willing? I've given a name to the future that the toilet revolutionaries are striving to bring about: Loo-topia. Eden had just one snake, but to get to Loo-topia, the world is going to need a whole lot of Sneeks.

What's in the Toilet?

Tinkle, plop, *kawoosh*! But wait, before you walk out of the bathroom and go on with your life, what exactly was it that you just flushed away? Let's start with poop and pee. These are the major ways your body rids itself of the stuff it doesn't want or no longer needs. There's considerable variation in what people produce. According to a paper from researchers at Cranfield University in the United Kingdom that reviewed the published research on the topic, people expel about four and a half ounces of poop per day—that's almost the weight of three golf balls. But some people poop a lot more: one healthy subject pooped out a whopping twenty-eight ounces per day, which is more than the weight of seventeen golf balls. Another healthy subject, on the other hand, produced less than two ounces per day, which is just about one golf ball. Most people pass all of their daily poop at one sitting—a hole in one, you might say—though it's not unusual to do it more than once per day or once every other day.

So what accounts for these differences in volume? Let's break poop down to find out. The biggest fraction of shit is actually water. A healthy poop is about three-quarters H_2O. That water keeps the matter flowing through your intestines. Should you excrete too little water, you have constipation; too much and you have, of course, diarrhea. The daily quantity of poop of people hospitalized with diarrhea weighs some five times the average.

If you take all of that water out and analyze the remaining dried-out poop, about a quarter to a half consists of bacteria, some of it dead and some of it still alive. For the most part, these bacteria are your friends; they live in your digestive system and help to extract energy from your food, support your immune responses, and even strengthen the integrity of your gut. Amazingly, according to some estimates, there are perhaps more than a hundred trillion microorganisms—bacteria, as well as other types called archaea and single-celled fungi—in the human gastrointestinal tract, making the colon one of the most densely populated and biologically diverse microbial habitats on earth. Some scientists refer to the combination of you and your bacterial microbiota as a "superorganism." But since this community of microbes keeps changing and renewing itself, your body must constantly rid itself of its members—and that's why they come out with your poop.

A small number of the microorganisms in your poop are emphatically *not* your friends, however. People infected with norovirus can shed more than a trillion copies of the virus in just one gram of feces (about the size of a peanut). You need only ingest a small amount of a sick person's poop to get infected with whatever viruses, bacteria, or parasites are in there. To be successful, toilet technologies *must* eliminate that threat.

Most of the rest of the non-water fraction is undigested food, the proportion of which depends heavily on diet, including hard-to-digest carbohydrates such as dietary fiber (about a quarter), some nitrogen-containing substances such as protein (up to about a quarter), and fats (up to about a sixth), for which your body doesn't have use. Dietary fiber—which, chemically speaking, consists of long chains of sugar molecules—is particularly significant: although fiber is hard for your body to break down, it is nonetheless important for healthy digestion (what grandmothers everywhere refer to as being "regular" and why they eat their bran). The size and weight of your poop, as well as the rest of its composition, depend a lot on how much fiber you consume, not only because of the fiber itself but also because it carries water with it. More fiber equals more water equals bigger poops. Because people who live in low-income countries tend to have more fiber in their diets, their poop weighs on average about twice as much as poop from people who live in high-income countries. Similarly, an average vegetarian's poop is more weighty than an average omnivore's poop.

On top of that, your poop contains rejected compounds such as calcium phosphate and iron phosphate, as well as body castoffs, such as debris from cells shed from the gut's mucous membrane, bile pigments (which make poop brown), and dead white blood cells. The primary odor of feces comes from an aptly named compound called skatole, a by-product of the digestion of tryptophan, an essential amino acid that you get from foods such as seeds, nuts, eggs, meat, and dairy. Depending on your diet and environment, your poop might also contain food colorants, medicines, and even microscopic plastic particles, which slough off of bottles and food packaging.

Moving on, urine is the waste product of the body's metabolic activity. Secreted by the kidneys, stored in the bladder, and then squirted out the urethra at a rate of about a third of a gallon per day, the yellow stuff is about 95 percent water. How much you pee depends largely on how much you drink. Sweating and exercising decrease your pee volume, as do age (older people pee more than younger ones) and race (Black women pee less than white women—go figure!). Interestingly, frequency of peeing isn't very well documented—one small study of seventeen adult women in the United States recorded eight urinations in a twenty-four-hour period. That sounds about right to me.

Aside from water, the most important component of urine is a nitrogen-rich crystalline compound called urea, which is what results when your body processes the protein that you eat. People who eat more meat will have more urea in their urine, along with more potassium and phosphorus compounds, making it a more potent resource for fertilizer. That phosphorus makes urine glow under a black light, but the yellow of urine has a different origin: some of the same bile pigments that make feces brown get reabsorbed by the bloodstream and then processed through the kidneys. As for microbes, researchers have found some of them in urine, including those that cause diseases, although many fewer than in poop. If you take medicine (antibiotics, antidepressants, and so on), those might come out in your urine, too.

Poop and pee aren't the only matter we flush down. Some flush menstrual discharges, which contain some blood from the arteries in the uterus but are largely, in fact, cells that build up in the lining of the uterus in anticipation of pregnancy. People vomit the contents of their stomachs into the toilet. And they also use it sometimes to get rid of spit and snot. The toilet is a willing and convenient receptacle for all of that bodily waste.

After we pee and poop, we clean ourselves and flush, often with clean, drinkable water. On top of that, when we scrub the toilet, the cleaning chemicals also end up down under. And people all over the world use toilets to discard what doesn't belong in there, from illicit drugs to baby wipes to unwanted animals, which can cause tremendous problems for sanitation systems and the environment. But, unless the police or sewer workers come knocking at the door, people just flush and forget. After all, history has taught us to think of the toilet as a trash can, when, as we'll see, it holds the potential to be a recycling bin.

A Very Short History of Toilet Systems

Like almost every aspect of the toilet, its past has been neglected, though archaeologists and historians have begun to remedy that. Let's start with a romp through that sparse history so we can better understand where we are and where we need to go. In the beginning, there was probably just a spot outside of camp or agricultural communities, in a field or wood or on a beach. When settlements got bigger, they came up with a patchwork of other arrangements, though those weren't sanitary in the modern sense. They handled, to greater and lesser extents, stench and mess, but they didn't fully separate communities from their collective poop.

Among them were the basic technologies of pipes, pits, and pots, which arose piecemeal, invented independently by many early cultures. In ancient Mesopotamia, the "cradle of civilization" where complex cities first grew, a settlement in modern-day Syria of the late fourth millennium BCE had clay pipes that probably carried toilet waste as well as rainwater and washing water, but those could have pooled into stinky, infectious messes where the pipes ended. By the early third millennium BCE, simple pit latrines appeared. While not celebrated like this region's other achievements, such as animal domestication, the wheel, and writing, many of these underground collection holes weren't too shabby, with linings of perforated ceramic rings that would have prevented collapse and allowed liquids to seep out into the surrounding ground, stopping them from overflowing. But this toilet type was uncommon so perhaps unappreciated; most people may have preferred an indoor pot or outside spot.

In the third millennium BCE, the city of Mohenjo-daro in modern-day Pakistan built the first comprehensive system of open sewer drains, available to all, which rainstorms flushed into soak pits, containing the waste. A few millennia later, ancient Romans made another enormous leap when they built the first large networks of underground sewers, which used gravity. The Romans' sewers mainly handled stormwater but also carried away some of the cities' pee and poop. Medieval Europeans, however, failed to maintain or expand these sewer systems, so they fell into disuse and disrepair in later centuries. Toilet alcoves called garderobes protruded from the sides of castles, allowing poop to plop into the moat. In some better-organized towns, "night-soil" workers collected waste, taking it to the countryside to fertilize farmland.

By the middle of the nineteenth century, London had gotten so big that it was drowning in poop. This era saw the popularization of flushing "water closets," which used piped-in water to clean the toilet bowl and carry away waste. The idea wasn't entirely new: an intriguing concept for the flush toilet appeared in the Knossos palace on the island of Crete as early as the second millennium BCE, during the Minoan era. The lavatory, in the residential quarter of the palace, was an indoor room containing a seat over a drain. Outside the door, a hole in the floor allowed someone to pour a jug of water into the drain, rinsing it out while, one imagines, avoiding any unnecessary backsplash.

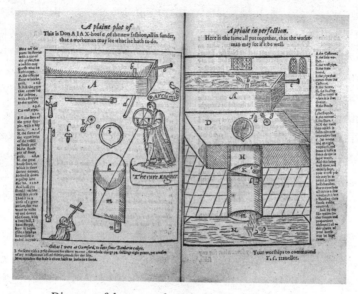

Diagram of the water closet in Sir John Harington's
The Metamorphosis of Ajax.

In the late sixteenth century, inventor Sir John Harington ushered in the modern era of the flush toilet. The biggest problem with privies at that time was that they stank, he wrote in his 1596 satirical book, *A New Discourse of a Stale Subject, Called the Metamorphosis of Ajax* (*Ajax* being a reference to both a legend from Ovid's *Metamorphoses* and "a jakes," the colloquial name for a privy at that time). "And for reformation of this, manie I doubt not, have ere this beaten their braines and strained verie hard, to have found out some remedie, but yet still I find all my good friendes houses greatly annoyed with it." Harington's "remedie"—a metamorphosed jakes—consisted of a cistern with water in it, which fed into an oval bowl that could be emptied by way of a screw-operated brass sluice on the bottom. When not in use, the bowl would contain a half foot of clean water that would prevent smells from wafting up from the pipe underneath it. At the time, doctors believed that bad odors could cause disease, and so Harington's invention, he thought, would not only smell sweeter but also prevent illness.

The water closet didn't catch on among the masses until after 1775, though, when a leading London watchmaker, Alexander Cumming (sometimes Cummings), took out the first patent for an "improved" version with a valve that slid open to release the water and waste. Just as the valve closed again, water filled in over it. Importantly, this invention also included a pipe shaped like a sideways *s* beneath the trap (this later became more of a *u*). Water settled in the dip of this *s*, which was even better at preventing stinky gases from wafting up. Three years later, inventor Joseph Bramah made another leap when he replaced the sliding valve, which tended to gather a crust, with a self-washing valve.

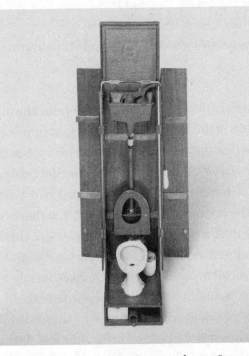

A model of George Jennings's water closet, 1895–1905.

The next century brought further refinements to this design, making it more hygienic and easier to use and taking advantage of the growing availability of piped water and sewers. In the mid-1800s, George Jennings became the next big name in toilets when he moved the basin and the trap closer together, discarding the valve and rendering the device so simple that he called one model the Monkey Closet. By about 1900, the WC had evolved into a ubiquitous single-piece ceramic self-cleaning mechanism called the wash-down closet and, ultimately, the toilet.

As they grew in popularity over the nineteenth century, toilets made life cleaner inside but nastier outside. Some of the sewage flowed out of toilets through disorganized gutters and sewers into the Thames. Some of it poured into overwhelmed cesspools, seeping into the groundwater or overflowing into yards. Cholera and typhoid outbreaks tore through the population, but doctors didn't understand the link to water contaminated with poop; they still thought that bad air caused sickness. During a heat wave in 1858, London began to experience what came to be known as the "Great Stink." Repulsive smells emanating from the festering Thames made the city practically unlivable and reportedly set citizens to fainting and sent legislators fleeing from Parliament. The lawmakers agreed to build a comprehensive sewer system to take away the foul stuff, setting an example followed by the rest of Europe, North America, and ultimately many other cities worldwide.

At first, sewers mostly just shifted the pollution downstream. It wasn't until the 1920s when the final piece of the modern sanitation puzzle, biological wastewater treatment, was put in place. Chemist and bacteriologist Gilbert John Fowler of Manchester, England, and his colleagues developed what's known as the "activated sludge" process, which harnesses naturally occurring microbes to clean the wastewater. In a typical activated sludge plant today, sewage flowing in first encounters screens and settling tanks, which remove large objects, grit, and a lot of the organic matter, including mushed-up poop. After that, the liquid travels to an aeration tank, where microbes, fed oxygen in air that's constantly pumped through the tank, break down the remaining organic matter in the liquid and clump into agglomerations called flocs, which then settle out in a clarifier tank. Since these flocs are rich with active microbes, some of them get recycled back into the aeration tank to help the process continue. What's left is clear water that's nearly—but not completely—free of pathogens; today, it may undergo at least some additional treatment to disinfect or filter it to remove nutrients and contaminants before flowing out into a nearby water body. The remaining sludge from the treatment process gets collected, processed, and disposed of—*how*, as we will see, depends on many factors.

This then became the "gold standard" in sanitation systems: your pee and poop, plus toilet paper, plop down into water. You flush, sending that mix into the sewer, which then flows to a wastewater treatment plant, which treats the sewage. Born in England, it has propagated to nearly every corner of the planet.

Some people stop the story there. But that's where our story really gets going, because the work is full of flaws. For one, though the gold standard has been around for a century, most places have not achieved it. Low- and middle-income regions, and even, to some extent, some areas in high-income countries such as the United States, resemble Victorian London, with a variety of toilet types, many unsafe, existing side by side, and with little oversight. Hundreds of millions of people worldwide still practice open defecation, meaning that they don't use any toilet. Open defecation is unavoidable for some because they don't have any facilities, though they would like them. Others do it due to custom and religious beliefs; in India, defecating inside the home was long taboo and many people considered a visit to the designated outdoor spot pleasant and social. On top of that, according to the best available data from the World Health Organization/UNICEF Joint Monitoring Program for Water Supply, Sanitation and Hygiene, as of 2017, about 700 million use toilets that nearly amount to open defecation—such as "hanging latrines," which are built over running water, or some "bucket toilets." More than 600 million people don't have private toilets in their homes but share them with surrounding households or neighborhoods.

Of the rest, nearly 2.2 billion people have "basic sanitation," which for the most part means pit latrines. *Basic* in this term sounds pretty okay, as if it meets a minimum level for safety. But it's not okay at all, really, especially in cities. In many cases, there's no safe way to empty and treat the waste in the latrines when they fill up, a situation that engineer and economist Elizabeth Tilley has dubbed a ticking "sludge bomb." Either the latrines overflow and go out of service, or someone—often an informal, unregulated, and unprotected worker—comes to empty them, at times just dumping the contents nearby. Septic tanks, which are underground containers that perform some microbial treatment, can also fall into this category if poorly installed or maintained—and many of the septic tanks in the United States fit that description, even if they're not officially recorded as such in the data.

Some sewer systems are included in that "basic sanitation" statistic. In most countries of sub-Saharan Africa and Latin America, as well as in many other lower-income countries worldwide, less than half of all wastewater flowing through sewers encounters an adequate treatment plant, or any treatment plant at all. As environmentalist Sunita Narain of the Centre for Science and Environment in New Delhi, India, writes, "Most of our rivers are today dead because of the domestic sewage load from cities. We have turned our surface water systems into open sewage drains." That leaves only 3.4 billion, less than half of the world's population, with what the experts now call safely managed sanitation, which can range from properly serviced pit latrines to comprehensive sewer and wastewater treatment systems.

These numbers actually represent progress. Still, despite the urgency of the problem, humanity isn't doing nearly enough to close the giant remaining gap. In 2010, the United Nations enshrined a human right to sanitation equal to the right to water. In 2015, the governments of the world agreed on Sustainable Development Goals.

The sixth goal compels them to "ensure availability and sustainable management of water and sanitation for all." Specifically, by 2030 they must "achieve access to adequate and equitable sanitation and hygiene for all and end open defecation, paying special attention to the needs of women and girls and those in vulnerable situations." But a 2019 World Bank report estimates that meeting this water-and-sanitation goal will cost low- and middle-income countries some $406 billion to $509 billion annually, equivalent to between 1.1 percent and 1.4 percent of their gross domestic product. That's three times current spending. Dishearteningly, the government spending on sanitation that does happen tends to go to relatively well-off urban areas with sewers because those are the projects that are the easiest to subsidize, as well as the residents with the most political power.

It's a frustrating situation, since government spending on sanitation makes good public health and financial sense. For every dollar invested in universal sanitation, according to one estimate, funders could expect an average return of $5.50, thanks to benefits such as reduced health-care costs and a more productive and involved workforce. To understand why, just consider that WaterAid estimates that women globally spend 97 billion hours per year looking for a safe place to relieve themselves, which amounts to more than all of the hours worked in a year in Germany, the largest economy in Europe.

Even in the United States, wastewater infrastructure—like so much of the nation's infrastructure—is aging and underfunded, risking heavy backsliding. In its 2017 Infrastructure Report Card, the American Society of Civil Engineers gave America's wastewater infrastructure a D+. "Federal funding, once the driving force behind water infrastructure development, has declined precipitously in recent decades, reducing the support available for communities to build and maintain water and wastewater systems," according to another report. American communities are largely left to foot the bills for wastewater treatment on their own, and many now have large backlogs of maintenance and upgrades that they can't afford. Most can't keep up with the latest innovations coming out of research laboratories.

And, in many ways, the gold standard is just not suited for many challenges of the future. Not the least of which is the major crisis of our times: a catastrophically heating climate, resulting in more, as well as more intense, natural calamities from fires to storms to heat waves to deep freezes. Wastewater treatment plants consume vast quantities of energy to pump and aerate sewage. In addition to any greenhouse gases associated with that power use, decomposing sewage emits them directly, including the particularly potent methane and nitrous oxide. One sad irony is that sanitation infrastructure is itself extremely vulnerable to the effects of climate change, like flooding and sea-level rise. And the people who live without good sanitation are also those most likely to suffer climate change's impacts.

Perhaps it's not surprising that a system designed in and for England a century ago and then exported as part of a kind of "sanitary imperialism"—a term applied approvingly in 1927 by Columbia University political scientist Parker Thomas Moon—is starting to look anachronistic. The world needs something new, but it's not at all clear *what*, since the sector doesn't have any plug-and-play solutions aside from the gold standard. That's changing, though. In recent years, alternatives have appeared as people propose new ideas not just for the toilet itself but also the rest of the system, which includes transportation, treatment, and reuse of the materials in the sewage. Some of the ideas build on the gold standard, while others hearken back to older technologies, like the night-soil systems abandoned in favor of sewers. Ultimately, the change has not only challenged notions of what progress looks like but also opened the door to a broader definition of sanitation.

The New Toilet Science

When environmental engineer Claire Furlong wanted to study whether a certain kind of worm, called the tiger worm, would eat fecal sludge, she was in a bind. While sewage is the icky mix of stuff, from pee and poop to water and trash and soap, that flows through sewers, fecal sludge is the mush in pit latrines. But in London, where Furlong was working, there hadn't been any pit latrines in recent memory. "That's why I shipped out of London to the wilds of Wales to collect hippie shit," she tells me. The staff at the Centre for Alternative Technology in Machynlleth gladly pooped in buckets with compostable liners, even referring to themselves as the Poo Club and wearing smiley-face badges. Twice a day, Furlong collected the donations, "homogenized" them by squishing or blending them all together, and fed the mixture to her worms—which, as it happens, ate it up, contributing to a new product called the Tiger Toilet.

Accelerating toilet innovation has required coming up with a remarkable variety of new supporting tools, techniques, and methods—a new toilet science. Getting the materials needed to test the technologies is just one of the problems, but it's a big one. Fresh feces aren't the same as fecal sludge, says Furlong, who now teaches non-sewered sanitation, but they served the purpose better than, say, the heavily diluted and lightly treated sludge collected at the first stage of a modern wastewater treatment plant. On the downside, it's not easy to find donors. One man at the Welsh center, she remembers, asked whether she could somehow identify his poop as belonging to him—perhaps worried that she would judge him on the amount, consistency, or contents. Another problem, she says, is that people tend to poop at home, either in the morning or at night, so a work-based collection point doesn't fill up as quickly as you might think. Most people aren't keen to tote their feces to work in a Tupperware container. And in some cases, experimenting with people's poop requires ethical clearance from review boards.

To overcome those challenges, there's fake urine, feces, fecal sludge, sewage, and even menstrual blood. When the Gates Foundation created a competition for radical new toilet concepts, it asked entrants to flush synthetic feces, made by a company called Maximum Performance, that are a combination of soybean paste and rice, extruded with a sausage machine. As Maximum Performance engineer John Koeller explained at the time, it "is the closest thing we have found to 'the real thing' both in weight as well as the way it begins to break apart when flushed."

But toilet innovators have found that that's not always the right fit for many of the laboratory tests they run for everything from how well poop moves through pipes to how it degrades in pit latrines to how much energy is stored in it. Researchers at Eawag, the Swiss Federal Institute of Aquatic Science and Technology in Dübendorf, reviewed the pros and cons of all the known poop "simulants." The paper, written in a dry, academic tone, can't help but produce chuckle after chuckle, as it refers to recipes that include red potato mash, peanut butter, pumpkin filling, brownie mix, wheat flour, salt, vinegar, miso paste, ground walnuts, and parts of pig and chicken intestine. You could almost make these fake poops in your home kitchen—and, indeed, the paper concludes with detailed instructions for a new synthetic poop that the authors devised. But all of the formulations have some drawbacks, for example: "the large amount of baker's yeast included in this recipe makes it physically very different from real human stool as it inflates like bread dough, and yields a sticky, unshapable slime." Ultimately, the paper's authors conclude, while synthetic versions have a role to play in research, there's no substitute for the real thing.

There are also new tools for understanding the risks communities face from poor sanitation. Public health engineer Barbara Evans and colleagues realized that the mountains of detailed reports that the sector produced weren't making a dent in policy. So, during a project with the World Bank that started in 2012, she has explained, "We decided to draw a picture for any given city of where all the waste comes from and where all the waste goes." Essentially, it's a series of horizontal bars: green ones that point from left to right represent the fecal waste flows that successfully reach treatment; red ones that take an alarming downward turn represent the flows that end up in the environment. Doing away with euphemisms, they called it the shit flow diagram.

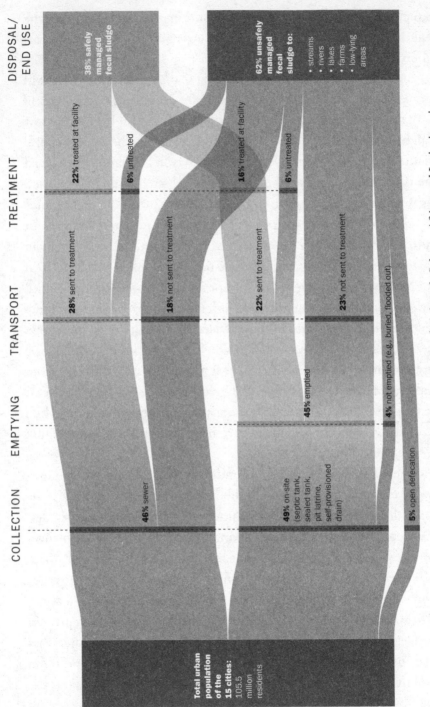

The combined flow of fecal waste in fifteen cities in South Asia, sub-Saharan Africa, and Latin America. An analysis by the World Resources Institute found that only 38 percent of the waste makes it safely through treatment.

I have seen quite a few of these by now, including one for Cap-Haïtien in Haiti, where I visited, and which faces problems typical of low-income cities. A team from the Inter-American Development Bank (IDB) did extensive research, including holding twenty meetings and twelve focus groups and also visiting more than fifty places in the city and surrounding areas. The diagram revealed that 99 percent of waste in Cap-Haïtien ends up, by various pathways, contaminating the environment. Without sewers, many residents rely on pit latrines and septic tanks, but there's no good place for the waste emptied from them to go: one rudimentary treatment site built in 2013 "was never completed and is not functional," the accompanying IDB report reads, although city officials claim it gets used from time to time. Unregulated and often stigmatized workers, called *bayakou*, who empty poor residents' pit latrines by hand, usually bury or dump the contents in nearby ditches or waterways. Vacuum trucks spread the waste from wealthier residents' septic tanks at informal sites, such as in an unsecured private field I visited, where people walk through and animals graze. Much of this pollution happens unnoticed, even by the residents themselves, but the upsetting red bars render the situation clear: 11 percent "open defecation," 44 percent "not contained," 42 percent "not delivered to treatment," and 2 percent delivered to treatment but "not treated." Just 1 percent of the waste gets safely treated, thanks to a program called SOIL (Sustainable Organic Integrated Livelihoods).

When testing toilets in the field, researchers face another problem—finding somewhere to do it. In Sneek, the original facility that served thirty-two houses is being resurrected as a new research site, part of the European Union–funded Run4Life program, which is developing technologies that recover nutrients from wastewater treatment systems at a total of four sites in Europe. As part of the project, the complex's residents have gotten the latest vacuum toilets—the same kind that I used—and the garage is getting an upgraded anaerobic digester. The collaboration includes not only universities but also fertilizer companies that will help the researchers understand how they can slot products into existing pipelines.

Many low-income communities don't relish being test subjects for well-meaning engineers from elite universities, and engineering teams often lack the expertise needed to handle those relationships. In Durban, South Africa, toilet innovators can take advantage of a program called the Engineering Field Testing Platform, which trials prototypes in the informal settlements around the city, which tend to have poor sanitation services, as well as in primary schools and rural households. The platform staff do the hard work of selecting the locations, fixing the equipment when it breaks, and communicating with the residents who will use the prototypes. "There's a lot of trust-building," Becky Sindall, the operations manager of the field testing platform, explains. "Imagine if . . . somebody comes along and strikes up a conversation with you about, so, how often do you go to the toilet, and what do you like about the toilet, and what do you not like about the toilets? You can start to get some funny looks."

The field tests can also help engineers understand how the toilets might be used in real life—like the time the team found foam bubbling out of a high-tech treatment unit that can both capture nutrients and treat water for reuse in toilet flushing. At first, they thought that used laundry water was making it into the system and foaming. But, upon further investigation, they realized that it was the toilet-cleaning solution that the caretaker was using. In South Africa, people connect foam with cleaning power, so the products foam more than elsewhere. The team had to empty, clean, and refill the unit, as well as train the caretaker on appropriate products for that model.

The new toilet science means bringing scientific rigor to a field that has long relied on conventional wisdom and whose practitioners often don't discuss failures outside of after-work drinks, says Dani Barrington, a sanitation engineer at the University of Western Australia. She and her colleagues, including Sindall, have created a movement encouraging water and sanitation professionals, especially those who work in development, to "talk openly when things go wrong," in order to avoid repeating mistakes. At one conference that I attended, Barrington, in a sparkly black-and-white dress with red pumps, hosted an audience-participation game show that highlighted a public toilet in Kenya that, because it didn't have connections to piped water or an adequate sewage system, got rented out as a church. Another was a campaign against public urination that involved spraying urinators with water, an intervention that might well work but could also do harm by physically hurting and humiliating people, as well as causing them to miss work due to soaked clothing. At the end, audience members shared their own stories; in one, organizers told children to blow whistles when they saw open defecation. That led to some of the children getting beaten up by the men they were calling out.

When she was in graduate school, says environmental engineer Linda Strande of Eawag, only a few universities in the United States had sanitation courses that focused on pits and tanks, and her colleagues found her interest in the topic "peculiar." Despite the fact that these forms of sanitation serve a majority of the world's population, including one in five Americans, most experts saw them as "temporary holes in the ground," which would eventually get replaced by sewers, rather than as technologies worthy of study and improvement. Although most research funding goes into conventional centralized systems, adds engineer Francis de los Reyes III, of North Carolina State University, "we can use the very best approaches to understand the problems [in low-resource contexts], and we should come up with solutions that are not just 'good enough,'" he says. "These problems deserve the best science."

The Transformed Toilet

The coming chapters will present many new concepts that could become part of a Loo-topian future, in which everyone has access to sustainable, healthy toilets. That no one concept has emerged as a winner has been a frustration for me at times because it makes it hard to know which ones to highlight—so many of them seem to be just waiting for a big break, like actors on the audition circuit. Some of those featured in this book will take off, while others will fizzle.

The Netherlands is a good case study in the barriers to transformation. In a nation of water engineers, the Dutch maintain high levels of support for the water sector, with the second-highest annual expenditures on water and sanitation per capita in Europe, after Luxembourg. Democratically elected Water Boards, which manage regional water systems, work toward a long-term goal of centralized wastewater treatment plants as "energy and resource factories" (in the United States, the comparable new euphemism is Water Resource Recovery Facilities). Water expertise is one of this country's main exports, and it boasts several top water engineering universities, as well as the IHE Delft Institute for Water Education, just a short train ride from my home, which is an international graduate school that, among other subjects, teaches future sanitation leaders how to handle the poop piling up in pit latrines. "Stupid infrastructure investment will lead us from crisis to crisis to crisis," the country's crusading Special Envoy for International Water Affairs, Henk Ovink, has said, referring to conventional water systems. "We need to become proactive." Still, some 99 percent of the population has a sewer connection—and, as far as I can tell, likes it—which means the country is one of the most "locked in" to centralized infrastructure. So while the overall environment is good for pilots like Sneek, it's less so for adopting radical change on a large scale.

In low-resource places, other barriers stand in the way. There may not be piped water, consistent electricity, or a pool of engineering talent. The area may be too crowded, as well as populated by people who move often, have little money, and command little political clout. But as I've talked to experts and entrepreneurs and also regular folks, read report upon report, and traveled to places both far-flung and close to home, I've found that the best solutions also have a lot in common. There are some things that any transformed toilet should do, if it will be better than the toilet systems we have today.

A transformed toilet will head off old and new health threats. It will not only thwart pathogens like those that cause cholera and typhoid but also protect against a modern scourge: a wide range of man-made pollutants, from antidepressants to microplastics, that enter our sanitation systems. Beyond that, it might even monitor the daily deposits of users, communicating with doctors and public health officials in order to catch individual diseases and community outbreaks early.

A transformed toilet will inexpensively, flexibly manage all of the waste in an area. A network of high-tech, off-grid toilets could do this. Or a small sewer system that serves one neighborhood, connected to a low-tech treatment plant. Or tanks or containers emptied by a service that treats waste in a composting center. Or even a mix of these.

A transformed toilet will not hog resources. Unlike with other kinds of waste, such as metal and plastic, we can't stop creating pee and poop. Pooping in itself is not a sign that you're anti-environment; it just means that you're alive. But, with changes to technology, behavior, and laws, we can reduce the amount of water and energy used by sanitation systems, as well as the amount of sludge that's treated as trash.

A transformed toilet will generate resources. The treatment of toilet waste can be a kind of manufacturing process, with excreta as the raw material. Products derived from that material are surprisingly diverse, including water, fertilizer, fuel, animal feed, fertility drugs, and components of bioplastics, bricks, and asphalt. Some sewage even contains gold and silver. The trick is mining all those riches.

A transformed toilet will have a financial plan. Selling recovered resources can create some financial stability for a sanitation system, so long as there's a market for them. But experts think that it's unlikely to cover most of the costs of safe sanitation. So sanitation services must gather together diverse sources of financial support—from the government above all—to provide continuous funding, or else they risk moving backward.

A transformed toilet will appeal to those who are meant to use it. All people come with cultural and psychological baggage around toilets, and that baggage can interfere with their willingness to accept new ideas. One way to appeal is to design a system that fits people's expectations. Another is to nudge people's expectations toward new ideas. Either way, people should have a say in designing and choosing the toilets that serve them.

A transformed toilet will take account of everyone in the community. That means all genders, all ages, and all backgrounds, each of which brings its own special challenges. That means people with disabilities and disease, visible and invisible. And it also means taking into account the often unsung or stigmatized people who clean, empty, and service toilets, whose work is usually unpleasant and often dangerous.

A transformed toilet will fit into local contexts. People need toilets everywhere they live: in deserts and rain forests, in cities and on farms, at music festivals and in refugee camps, in economic boom times and after disasters, in functioning democracies and corrupt dictatorships, and maybe someday on Mars colonies. The worst idea, it turns out, is to impose a solution that worked in one place without taking a good look around in the new place. This can go the other way, too: sometimes the best answer is to improve existing infrastructure rather than try to replace it.

We shouldn't settle for the toilets we've inherited. The emerging toilet paradigm will offer a kaleidoscope of toilet systems, geared for different environments, different preferences, and different values. Ultimately, whoever you are, your toilet shouldn't be just be a default— it should be right for *you*.

Paging Dr. Toilet

It's always on call.

The world is divided up into two kinds of people—those who look at their body waste in the toilet bowl, and those who don't.

—John Gregory Dunne,
Nothing Lost (2004)

It Takes Guts

Vik Kashyap wants to image your poop. A Silicon Valley entrepreneur, he came to the idea after his relationship with his toilet changed for the worse. In 2003, at age twenty-eight, he was living in New York City when he developed ulcerative colitis, a chronic autoimmune condition. "It was awful," he says. "Bleeding and weight loss, going to the bathroom fifteen times a day, very unpleasant." The usual drugs didn't work, and he faced a surgery that would remove his colon. Unwilling to accept that outcome, he attempted to hack his gut. Guessing that his gut microbiome might be messed up, he tried transplanting fecal bacteria from a healthy person into himself. It didn't help.

Then he landed on what's sometimes called the hygiene hypothesis, which lays some of the blame for an uptick in many autoimmune diseases at the foot of, well, hygiene. While up to a tenth of the population in developing countries carry an intestinal worm infection, elsewhere the infections are rare. Without these natural outside threats to respond to, the hypothesis goes, the immune system turns on healthy tissue instead, leading to conditions such as multiple sclerosis and lupus as well as ulcerative colitis and Crohn's.

American researchers wouldn't accept the legal liability for what Kashyap was about to do, so he turned to a parasitologist in Thailand. Together, and with difficulty, they bred a parasite called *Trichuris trichiura*, or the whipworm, which Kashyap sourced from the poop of an eleven-year-old girl there. In much of the world, people get it when they ingest its eggs. The eggs then hatch in the small intestine, releasing larvae that grow to about an inch and a half in length (about the width of two adult fingers). Attaching to the colon, the females shed between three thousand and twenty thousand eggs per day for about a year, and these exit the body with stool, starting the cycle again. Whipworms can cause hosts to poop a painful mixture of mucus, water, and blood or suffer from anemia, a lack of iron. But often they cause no disease at all.

Kashyap felt he had nothing to lose. He ate the eggs. It took several tries, but, to his amazement, it finally worked. After a few months, his ulcerative colitis symptoms went into remission. Later, he convinced immunologists at the University of California–San Francisco to study him as he cycled through two flare-ups, which he believes occurred as the worms aged and died, requiring him to reinfect himself—this time with worms he isolated from his own poop. They published their findings in the scientific journal *Science Translational Medicine* in 2010.

Naturally, he was relieved, but he also wanted to know how to stay in remission and possibly feel even better, without continuing to dose himself with worms. Using fitness trackers and other technology, he gathered as much data as he could on himself in the hopes that it would tell him how his everyday choices of food and nutritional supplements affected his health, he says. "I kind of became obsessed, actually, with figuring out a way to do this."

At the same time, he needed a new professional direction. Amidst the anguish of illness and release of remission, he had discovered that two of his co-founders at his health-care company had committed fraud and embezzlement involving millions of dollars, leading to long jail sentences for them, as well as his own resignation. All paths—the personal and the professional—converged on the toilet, which he hoped to transform into the ultimate health device, for himself and for others.

You may not think about cholera every time you flush, but concerns about health, as well as about comfort and convenience, have influenced the way our toilets look today and how we use them. A good shit is, according to many, one of the most satisfying experiences known to humankind, and it's also key to good health. But the products of digestion can be dangerous if said shitter is infected. Poop, one could say, is the original hazardous waste. Experts categorize what writer Rose George calls "shit-related diseases" in a variety of ways. Sometimes it's by the route of transmission: water, food, or soil. Sometimes it's by symptom: diarrhea, fever, inflammation. And sometimes it's by type of pathogen: bacterium, virus, or parasite.

Before the era of modern toilet systems, terrifying outbreaks of these diseases were commonplace. On top of that, endemic levels of shit-related diseases were responsible for a background level of misery and death. Given these diseases' dramatic effects, our species may have evolved some hygienic behaviors against them long before scientists discovered pathogens. Even most animals have a hygiene instinct, according to the late behavioral scientist Val Curtis, of the London School of Hygiene & Tropical Medicine and the author of *Don't Look, Don't Touch, Don't Eat: The Science Behind Revulsion*. Bees defecate away from their colonies and raccoons use latrine sites. In humans, Curtis argued, this instinct became disgust. In the early nineteenth century, Darwin himself had already noticed that the emotion seems to transcend culture. And when Curtis asked people from around the world what they thought was disgusting and compared it to a medical manual, she concluded that "all of the things that people find revolting seemed to have some sort of role to play in the transmission of infectious disease," including bodily wastes, sick people, and spoiled foods.

In ancient societies, as people transitioned into sedentary agglomerations, which grew bigger and more complex, sanitation-related diseases may have found more fertile ground than in hunter-gatherer groups. At the same time, humans' instinctive disgust may have propelled cultural practices such as latrine use and ritual handwashing and beliefs that in some cases also linked toilets to illnesses—although not necessarily the correct ones. In the Hebrew Bible, Deuteronomy instructs that encampments should have a designated area away from the camp for defecation and that people should dig a hole with a trowel because excrement is "indecent." Later, one Jewish sect upped defecation's status to ritually impure, even refraining from the act on the Sabbath (to the extent possible, one assumes). Ancient Babylonians recognized a low-ranking privy demon called Šulak, a "lurker" that could be responsible for bad luck, injury, or illness. Romans also feared demons that lived in toilets; to counteract them, they decorated toilet rooms with images of the goddess Fortuna, spells, and funny drawings that would inspire laughter, which was thought to ward off evil. In Japanese tradition, where toilets were protected by a *kawaya kami*, or toilet god, a dirty toilet was thought to lead to the birth of ugly and unhappy children.

These practices and beliefs wouldn't have been at all sufficient to prevent the spread of disease, since they didn't necessarily ensure the most crucial aspect of good sanitation—the separation of communities from their collective feces. In ancient Rome, despite the wide availability of toilets and sewers, infection by poop-borne parasites such as worms didn't fall by all that much, probably because, not knowing better, the Romans dumped sewage in rivers and spread poop on fields. "I think Romans were making progress towards thinking about ways to divert dirt and filth from living centers in their cities, but there was no such thing as a sanitary engineering system in their minds," classical archaeologist Ann Olga Koloski-Ostrow of Brandeis University says. Some historical figures even promoted excrement as healthful: the chronically constipated Martin Luther is rumored, possibly slanderously, to have eaten a spoonful of his own poop every day, a practice that isn't as dangerous as eating others' poop, since you're already infected by any pathogens in your own. It is, however, gross.

Among the many historical shit-related killers was typhoid fever, caused by the bacterium *Salmonella typhi*, which some historians have hypothesized may have nearly felled Caesar Augustus in the first century BCE, contributed to the deaths of 50 of the original 104 colonists at Jamestown, Virginia, in the seventeenth century, and sickened one-fifth of the American army in the Spanish-American War in 1898. (Typhoid fever is not related to typhus, which is spread via fleas and the like, although doctors once thought the two diseases were the same, thus the similar name. On the other hand, *Salmonella typhi* is related to the bacterium that causes salmonella food poisoning.) Once in the small intestine, the *typhi* bacteria use the body's defenses against it, multiplying within the white cells that the immune system sends to attack it and entering the bloodstream. The infection tends to cause prolonged fever as well as fatigue, headache, nausea, belly pain, constipation or diarrhea, and possibly a rash. Left untreated, the bacteria can travel to other areas of the body, damaging organs and tissues and causing internal bleeding, infections of the bone, or intestinal perforation, a splitting of a section of the digestive tract, which spreads the infection to nearby tissue, with often fatal results.

In an 1828 etching by William Heath, a fashionable woman
drops her teacup in shock after a microscope reveals that a drop
of Thames water teems with tiny "monsters."

As bad as typhoid was (and still is), however, it was outbreaks
of cholera that propelled key developments in sanitation. When
the disease broke out in London four times, starting in 1831, most
doctors believed in the "miasmatic theory"—that is, that you could
catch illnesses from bad, smelly air. In reality, as a result of increased
international trade and travel, the comma-shaped bacterium *Vibrio
cholerae* had begun to spread across the world from its reservoir in
the Ganges Delta of Bangladesh and India, riding on the surfaces of
tiny water-dwelling creatures called copepods. The origin of the term
cholera isn't known with certainty, but it may come from the Greek
word for "spout" or "gutter," which describes what the bacteria make
of the gut: once inside, they secrete a toxin that forces the intestine
to release water and salts rather than absorb them. If people rapidly
lose 10 to 15 percent of their body weight through this diarrhea, they
won't have enough fluid in their bloodstream to keep up their blood
pressure, which means their blood can't deliver oxygen and nutrients
to critical organs. This severe dehydration kills, even within hours.

In the mid-nineteenth century, a physician named John Snow began to doubt the miasmatic theory, proposing something much closer to the truth: that cholera spread when people ingested substances, particularly water, contaminated with the waste of infected people. During London's third cholera epidemic of 1853 to 1854, in which more than ten thousand people died, he observed that the disease struck a neighborhood near his practice particularly hard, and traced the origin to a sewage-contaminated well on Broad Street in Soho. He convinced officials to remove the handle on the well so that people couldn't use it, ending the outbreak. Yet it still took more than a decade—and more cholera deaths—before Snow's theory overturned the entrenched miasmatic beliefs. In the 1880s, discoveries of disease-causing microorganisms by German bacteriologist Robert Koch and French scientist Louis Pasteur established the germ theory of disease and inaugurated a new era of medicine. Koch himself isolated the cholera bacterium in 1883 in Egypt and India. Seeing images of the pathogens helped people accept the new germ theory, Curtis and her colleagues argued, because "the notion of an invading parasitic life form is so exquisitely disgusting." It didn't hurt that those life-forms lived in raw sewage.

The insight about germs had an effect on bathrooms. Early chamber pots and water closets were often encased in wooden boxes, which were sometimes decorated—as books, for example—to blend in with the rooms where they sat. The first ceramic toilets, installed in bathrooms, were likewise adorned, often with curlicued nautical themes as a nod to the water flowing through them. But at the turn of the century, as people got attuned to germs, they began to worry that the decorations could hide dirt, and so the simple white ceramic design, with as few nooks as possible, took over.

It also helped accelerate the expansion of sewerage and drinking-water purification. The effects went far beyond ending cholera and typhoid outbreaks. At the turn of the twentieth century, engineer Hiram F. Mills and health official J. J. Reincke independently discovered a relationship that then got named after them, the Mills-Reincke phenomenon. Tracking the cause of deaths in America and Europe, they found that for every death from typhoid fever that was avoided by improved drinking water, two to three deaths from other diseases, such as heart disease, tuberculosis, and pneumonia, were also prevented. The reason is still a matter of speculation. Drinking contaminated water may leave people's immune systems weak and unable to fight off other diseases.

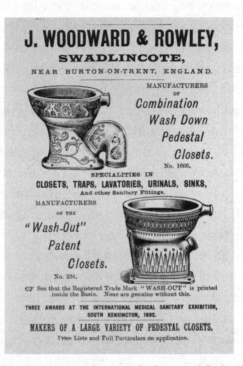

An 1894 advertisement for ornately decorated flush toilets,
known as "wash-down" or "wash-out" closets.

Modern sanitation isn't perfect protection, however. The flush of a toilet can fling pathogens throughout a room, potentially making them airborne and contaminating surfaces that others will touch (keep the lid down!). Dirty, poorly ventilated public toilets can be vectors for disease—though it's usually not the toilet seat that's the problem (wash your hands well and don't touch the door handle with your bare hands!). In the 2003 outbreak of the novel disease known as SARS-CoV in Hong Kong, faulty piping caused a residential building's plumbing system to become a vector for the virus to spread to 320 residents, leading to 42 deaths (don't let your drain traps dry out!). And, as Kashyap learned, scientists also think that there can just be too much of a good thing, such that living in highly hygienic environments can cause people's immune systems to turn on themselves.

Still, a safe toilet is like a "super vaccine" (and a painless one at that), the Gates Foundation's Doulaye Koné has said. And the era of SARS-CoV-2 (COVID-19), which caused the 2020 pandemic, has given us an even better analogy. The best available first response for the pandemic was *social distancing*: working from home, keeping six feet apart, and not gathering in large groups. For shit-related diseases, however, the solution is much easier. Using safe toilets *is* social distancing—from poop.

Lost in Translation

If you travel, you may have noticed that the concept of sanitation allows for some, um, variation in the user interface. In truth, each culture has its own beliefs about the healthiest way to poop. Take the great sit-versus-squat debate. In ancient Mesopotamia, archaeologists have found both pedestal and squat toilets, but today, the cultures have diverged, with squatting winning out in much of the world. I, as an American, come from a sitting culture but have squatted in many locations, from a municipal building in Japan to a zoo in Italy to a remote roadside in Alaska (oh, the places the mosquitoes did go!). But I never saw any other advantage to the posture—until after I had a baby. After pushing an infant's head through one's pelvis, it somehow becomes very difficult to squeeze out a simple turd—and, it turns out, squatting can help. I had a very good (and painful) laugh with my postpartum nurse, who told me that, for many of the women she cares for, passing the first bowel movement is almost like giving birth a second time.

In *Gut: The Inside Story of Our Body's Most Underrated Organ*, German physician-scientist-writer Giulia Enders makes an analogy between your gut and a garden hose, though maybe a bubble-tea straw is better. When you're standing or even sitting, your lower bowel—the hose or straw—has a substantial kink in it, which (usually helpfully) holds back poop. But when you squat, the canal straightens and the tapioca pearls can flow freely. She refers to one experiment in which twenty-eight subjects took less than half the time to defecate when squatting than on a "normal" toilet, and they felt more satisfied, too. A second experiment x-rayed volunteers who had eaten luminous substances as they defecated in various positions. "Squatting does indeed lead to a nice, straight intestinal tract, allowing for a direct, easy exit," she writes. Some even worry that the pedestal toilet has caused health problems ranging from constipation to hemorrhoids in the Western population. Certainly, doctors warn people from sitting for long periods on the toilet, so if you're reading this book on the loo right now, I'll take it as a compliment, but please stop.

In his classic 1966 study on the bathroom, architect Alexander Kira called the toilet "the most ill-suited fixture ever designed," but, today, to our rescue come the manufacturers of "defecation postural modification devices," low footrests that raise the feet so that sitting becomes more like squatting. The most well known of these is the Squatty Potty. But do they work? The evidence is slowly building in favor of them. Recently, a study at Ohio State University asked a sample of 147 healthy medical residents and their significant others to poop without a Squatty Potty for two weeks and then with a Squatty Potty for another two weeks, filling out surveys the whole time. The authors concluded that, yes, the device "positively influenced defecatory time, straining, and complete evacuation of bowels," at least according to the volunteers' subjective judgments. For the nearly 8 million people per year who are treated for constipation, devices like the Squatty Potty could be a "low-cost noninvasive option" for treatment, the authors write. Indeed, it's a no-cost option, as I myself discovered post-birth. Just find any footstool, put your feet up, lean a bit forward (what some have called *The Thinker* Position), and relax!

For the squeamish, squatting is probably somewhat more hygienic than sitting, since the user has less direct contact with surfaces, though squatting toilets can lead to more splash-back. Public toilets are no more contaminated than your home, according to research, since fecal bacteria tend to die off quickly in public restrooms, replaced by other microbes that are found on people's skin cells. As for the knees, it's true that some people in squatting cultures can't continue as they get older. But getting on and off the toilet is difficult for the elderly and mobility challenged in sitting cultures, too, which is why some inventors are working on toilets that can raise and lower to help people use them safely.

Squatting cultures also tend to be washing cultures. If you're used to toilet paper, it's hard to imagine life without it, hence the hoarding during times of crisis. In much of the world, however, you won't find toilet paper in the restroom. Instead, people clean themselves with water. The mechanics of this vary: people splash themselves with a jug or ladle with the help of their left hand (which then may not be used for eating—a problem for lefties like me), spray themselves with a handheld hose affectionately called a bum gun, and, in the case of the high-end robo-toilets usually associated with Japan, press buttons to activate a jet of water and a dryer. People who grow up with these methods find them second nature; newbies, however, often end up embarrassingly wet.

Marketers of toilet paper and bidets make bold claims as to their health benefits. Around 1900 Scott Paper advertised that their product could help stop dysentery, typhoid, and cholera, an overpromise at best. More recently, American bidet maker Tushy coined the term *gross butt syndrome* to describe conditions ranging from "hemorrhoids, UTIs, [to] general germy stank." In any case, adherents to either tradition tend to find the other horrifyingly unhygienic. How can you get yourself clean with just water? How can you get yourself clean with just paper? There is limited literature that addresses this question. Dr. John Swartzberg, a clinical professor emeritus at the University of California–Berkeley, writes in the *Berkeley Wellness Letter* that bidets could help with hemorrhoids, anal fissures, and pruritus ani (itchy anus), since wiping can aggravate those conditions. On the other hand, he writes, studies have found that the spray from a bidet can irritate the anus and affect the vaginal microbiome, and there's also a risk of damage from high pressures and scalding temperatures. On public toilets, especially in hospitals, improperly disinfected bidets could help spread drug-resistant bacteria, according to a study from Japan.

As for whether washing or wiping removes more fecal matter and germs, it may not matter in the end. As Shannon Palus, who reviewed bidets for The Wirecutter, puts it: "Your butt is probably fine with whatever level of microbes and poop it's sporting now. We do not have some national health crisis in America where our butts are too dirty. 'General germy stank' is not a real health problem."

Some are obsessed not with getting rid of fecal matter so much as holding on to it. In the more fecal-retentive parts of Europe, especially Germany, you will find a low shelf in many toilets, though these have been growing scarcer in recent years. The poop lands there first before getting washed into the water. These *Flachspüler*, which translates literally as something like "shallow flushers," serve two purposes. First, in this system, a big turd won't cause the toilet water to splash up onto your bum (one toilet manufacturer's website touts this convenience in beautiful German as *geringere Spritzgefahr*, or "reduced spray danger").

But the other is that the shelf makes it easier to inspect your stool before flushing it (termed *medizinische Stuhlkontrolle*, or "medical stool inspection"). One feature that seasoned stool inspectors look for is consistency. Gastroenterologists often ask patients to evaluate their bowel movements according to the seven-point Bristol Stool Scale, which roughly tracks the amount of water in the stool. A very watery stool—diarrhea—ranks as a 7, while hard, nut-like pellets are a 1 and reflect constipation. The ideal is a sausage-like 3 or 4. Though it may sound subjective, a study found that people are pretty good at rating their own poop.

Color can also be telling. Healthy poop is brown because it contains bilirubin, a pigment that gets excreted in bile from the liver, which turns from yellow-green to brown as it undergoes chemical changes during its journey through the digestive system. If it varies from that, as Enders explains in *Gut*, you might have a problem. If it is always light brown to yellow, you might have a largely harmless condition called Gilbert's Syndrome, in which an enzyme involved in breaking down blood works inefficiently. People who have it should avoid acetaminophen (Tylenol) since they don't tolerate it as well. A gray tint to your poop means a blockage between the liver and the gut, and black or dark red poop means internal bleeding—a reason to visit the doctor right away.

While most poop doesn't smell great, particularly foul smells emerging from the toilet can indicate malabsorption due to conditions ranging from celiac disease to inflammatory bowel disease to food allergies, as well as intestinal infections. The odors could also be benign, resulting from a change of diet or new medications. Interestingly, an infection with the bacterium *Clostridioides difficile* can sometimes make poop smell sickeningly sweet.

It's easy to make fun of the German obsession with *Scheisse*, or shit. But people around the world have surely been examining their poop for at least centuries and probably since the dawn of time. Many medicine traditions encourage examining one's poop for its diagnostic value. And, according to a study of volunteers in England, "most people admitted that they sometimes inspected their stools, men slightly more so than women"—92.8 percent and 89.2 percent, respectively. So, for all of our differences behind the restroom door, in this practice we may have found the one thing that everyone can all (secretly) agree on.

A Crying Shame

In 1999, Kamal Kar, a livestock specialist from Calcutta (later renamed Kolkata), India, went to evaluate a sanitation program in neighboring Bangladesh. Frustratingly, giving toilets to rural people, either directly or through subsidies, wasn't working. People with the donated toilets would continue to walk to informally designated areas, such as fields and railroad tracks. They would sometimes repurpose the toilet buildings as storage or animal sheds. As a result, disease would continue to spread.

As he tells it, Kar wanted to know what was at the heart of this problem. He went to a small community and asked residents to take him on a tour, showing him points of interest: not only where they defecated but also where they bathed, washed clothes, and collected their drinking water. These were just open-ended conversations, but "when people walked along with us, the realization started building up," he said in one of many recountings of this story. "Women started speaking and some pointing. And even closing their nose and mouth and all that." Through this process, people in the village came to understand that they were bathing and washing in water full of poop. "It's a terrible thing we have been doing here," they told Kar.

When he told the people that he didn't have money to build them toilets—it was just a learning exercise—something happened that astounded him: unwilling to go on as they were, the villagers themselves came up with the money. In that moment, he had "a flash of inspiration." People didn't need money to adopt toilets; they needed a shift in their mind-set. He developed a method that has since been deployed to millions worldwide, even as it has come under heavy criticism. He called it Community-Led Total Sanitation, or CLTS.

The method goes like this, though it can be altered for different environments: Facilitators visit remote communities, ideally ones where people are essentially naïve to the importance of toilets and the dangers of open defecation. The residents take the facilitators on a "defecation area transect walk" of the communities, during which the facilitators ask probing questions related to open defecation, like which family uses which area and what people do at night or when they are sick with diarrhea. The facilitators also draw attention to flies landing on or chickens pecking at shit on the ground. Although people are familiar with these sights, according to one training manual co-authored by Kar, "the embarrassment experienced during this 'walk of disgust' can result in an immediate desire to stop open defecation and get rid of these areas."

Later, during a community meeting, the facilitators teach another lesson, also in a Socratic style. Holding a clean glass of water, they ask if people would be willing to drink it. (Yes.) Then they take a strand of hair from their heads, touch it to a piece of shit, and put it in the water. Will people drink it now? (No.) Then facilitators point to the fact that flies have six legs and move between shit and food. What, then, are people eating along with their food? (Shit.) This exercise can generate an "ignition moment," also called triggering, in which people realize that open defecation results in everyone eating one another's shit. The solution—to the health risks as well as the feelings of shame and disgust—is then obvious to all: end open defecation and install toilets.

Due in part to the persistence of open defecation and unsafe toilets, today a large portion of the world continues to suffer from preventable and treatable shit-related diseases, both endemic and epidemic. These often disproportionately affect the very young, who weaken more readily from the dehydration of diarrhea. Globally, according to a 2019 study, diarrheal diseases caused by inadequate sanitation lead to 432,000 deaths and many more illnesses each year. When combined with inadequate water and hygiene, this "cluster" of risk factors causes some 829,000 deaths annually, of which 297,000 are children under five.

Even if children fight off the barrage of illnesses associated with poor sanitation, the cumulative effect of them can cause devastating harm through stunting, a condition that affects about a fifth of children under five worldwide. Stunting is not just about height: it is associated with poor cognitive function and low adult wages, among many other lifelong problems. Stunted mothers are more likely to have stunted children, which perpetuates the cycle of poverty, both of individual families and of entire nations, across generations. It's a condition of malnutrition, but it can also appear when there's plenty of food but a dearth of safe toilets and water. One reason is that, during bouts of diarrhea, a child won't eat and nutrients get flushed out of the body. Another is that parasites that colonize children's intestines steal nutrients for their own growth, leaving the children with less. Beyond that, multiple infections can cause some children to develop *environmental enteric dysfunction*, a still poorly understood state of intestinal inflammation. It's subclinical, meaning that it doesn't cause acute symptoms that would lead parents to seek medical care, but it still could disturb the gut's ability to absorb and deliver nutrients to the body.

Poor sanitation can also contribute to the problem of antibiotic-resistant pathogens, the rise of which could kill 10 million people each year by 2050, according to the United Nations, which calls the problem a global health emergency. When we fight pathogens with drugs instead of toilets, widespread illness from preventable diseases leads to a vast overuse of antibiotics, which can drive new antibiotic-resistant strains. On top of that, the antibiotics, excreted in our pee as well as flushed, make it into waterways when sewage gets dumped or treatment plants don't have the technology to remove them. These drugs could drive pathogens as well as naturally occurring bacteria to evolve resistance to them. In a study presented in 2019, the largest of its kind, researchers found common antibiotics in nearly two-thirds of test sites in seventy-two countries; about 15 percent of those exceeded safe levels, sometimes by more than three hundred times. The worst cases were in Africa and Asia. (The researchers found that animal waste and drug manufacturing were also sources of antibiotics in the rivers, in addition to sewage.)

Poor sanitation also affects health in an uncomfortable litany of ways unrelated to the pathogens in poop. Broken sanitation systems can leave dirty, standing water around, providing an ideal breeding ground for insects that spread diseases such as malaria, lymphatic filariasis, West Nile virus, and trachoma. People, especially children, sometimes fall into poorly built pit latrines, where they can drown. Without private toilets, people can be at higher risk of assault when they go out to relieve themselves, and a lack of toilets can contribute to mental health issues such as anxiety. People may stay away from health-care centers and schools that don't have adequate sanitation, contributing to a cycle of illness and poverty. Holding in pee and poop may cause people to suffer from urinary tract infections and constipation.

Community-Led Total Sanitation, or CLTS for short, deliberately harnesses the emotion of disgust to end open defecation. But critics argue that the approach can backfire, as it can also summon shame. Shame *can* be a helpful emotion if we use it, introspectively, to reform our ways, to become better people. But it can also be a toxic one, tied to a variety of mental disorders, that makes us want to "sink into the ground," writes psychologist Nick Haslam in *Psychology in the Bathroom*. Disgust's typical facial expression is a wrinkled nose, which blocks the nostrils; shame's is downturned eyes and a drooping head. "People feel ashamed when someone is disgusted with them and feel disgusted with others when they behave shamefully," Haslam writes.

In CLTS interventions, the conflation of instinctive revulsion and moral judgments about a person's worth can lead to bullying and stigmatization—and sometimes worse, write medical anthropologists Alexandra Brewis and Amber Wutich in *Lazy, Crazy, and Disgusting: Stigma and the Undoing of Global Health*. In one small Bangladeshi farming and fishing community, local sanitation committees took away pensions and work carts from poor neighbors who had not built toilets, and declared that the legal system wouldn't protect girls who were raped while defecating outside. Even if a CLTS intervention doesn't work—a community isn't successfully ignited or if it is ignited but doesn't manage to organize a response—members may grow more suspicious of one another or the government, reducing social cohesion and getting in the way of other initiatives. People may generally feel more ashamed and humiliated, now more aware of their problem but unable to correct course. CLTS proponents, for their part, acknowledge the potential for pitfalls but argue that the method works well when there are properly trained facilitators.

CLTS gives participants the freedom to choose the type of toilet they build, and they have often opted for subpar products, sometimes because they can't afford better ones or can't access high-quality labor and materials. Within a framework known as the "sanitation ladder," this isn't necessarily bad, since the idea is that people will willingly climb the ladder, replacing poor toilets with better and better ones, as they grow to appreciate them more and as their means improve. But that framework isn't always realistic: many people may not be able to accomplish that climb, and communities often revert to open defecation if their poorly built toilets fill up or cave in. They might even like toilets *less* than they did before. In Micronesia, CLTS prompted people to build toilets instead of defecating on the beach. But, in the absence of latrine-building experts, people built cesspits that polluted the main freshwater source under the atoll. So people abandoned the cesspits but didn't want to go back to their old practice because of shame, so they turned to hiding in bushes. While once the tide would wash the feces out to sea, after this fiasco it just lingered on the land instead, creating a greater risk than ever.

At a more basic level, critics have questioned some underlying assumptions of CLTS: that rural villages are naïve and homogenous and that people fail to use toilets out of ignorance instead of having perfectly rational reasons for doing so—including that the toilets that they've been given are disgusting or dangerous. Ultimately, it seems that CLTS has become a victim of its own success, so seductive for its supposedly quick action and low cost that donors and governments rolled it out on a massive scale before researchers could grasp how, or if, it was working and what its downsides might be. Despite its wide use, the evidence to support it is thin. As a trio of scientists wrote in 2019: "If clinical trials of a new pharmaceutical drug had results like those currently available for CLTS, the U.S. Food and Drug Administration would not approve the drug." When you add up all the costs, it is often not even that cheap.

So what does work? Although experts agree that building toilets *should* improve health outcomes, proving it has been surprisingly difficult. Recently three high-quality, randomized controlled trials in rural areas of Kenya, Bangladesh, and Zimbabwe showed little to no improvement in childhood diarrhea following "basic" interventions that decreased open defecation and increased the use of good pit latrines, clean drinking water, and handwashing. And the trials showed that the interventions had no effect on child stunting.

To explain the disappointing outcome, experts pointed to several factors. Primarily, because the studied interventions didn't reach everyone in a community, there was still quite a lot of fecal contamination around. One estimate is that three-quarters of a community have to use a basic toilet before it will glean most of the health benefits, and only 45 percent of people worldwide live in communities that meet that standard. Plus, a quarter of people worldwide do not have a good handwashing facility and only about one in four "potential faecal contacts" gets followed with handwashing with soap.

The results of interventions may also take a long time to kick in, as they did in Europe and North America. Stunting is a multigenerational problem, passed from mother to child. And some infections are particularly sneaky: think of poor Mary Mallon, known as Typhoid Mary, an Irish-American cook who spread typhoid fever to at least fifty-one people, killing three, over her career in New York in the early twentieth century, continuing for years after improved sanitation made the disease rare there. She had the misfortune of being an asymptomatic chronic carrier of the bacterium, possibly due to its ability to create a hideout from the immune system in bile and gallstones. Forcibly quarantined by public health authorities, she spent nearly three decades in isolation.

Some public health experts have taken the weak evidence for the health benefits of basic interventions to mean that money might be better spent elsewhere. Others argue that sanitation has benefits beyond its effects on childhood illness and that interventions should be even *more* ambitious, freeing communities from fecal contamination with better than just "basic" infrastructure. As the global haves develop innovations that will harness their toilet systems to do even more for health, the great moral risk is that the gap between them and have-nots will widen to more extraordinary proportions than ever before—even as some programs teach the have-nots to be ashamed of their situation. That, more than anything, would be truly disgusting.

Test the Waters

Meet Mr. Box—or so some of his handlers call him—a portable briefcase-sized bot that likes to hang out in manholes. Suspended with a wire cable over a sewer stream, the high-tech sampling device pumps wastewater through a series of filters that remove gunk and microbes and capture chemical compounds related to opioid drugs; then the filters get sent to the chemistry laboratory of the company Biobot Analytics for analysis. Scientists turn the data into maps and charts of drug concentrations throughout the city, returning these tools to the city's public health officials.

As receptacles for nearly all of the human waste in some cities, sewers can provide what's essentially a urine and stool sample for a whole population. Drugs that people excrete are usually a nasty problem for treatment plants: in downstream fish, painkillers can damage tissue, contraceptives can change sex traits, and antidepressants can alter behavior. But the idea of using drug residues for good—to monitor the health of a city—dates back to at least 2001, when environmental chemist Christian Daughton, then of the Environmental Protection Agency (now retired), proposed that wastewater could contain useful information about illegal drug use. Analyzing sewage could be "the first feasible approach to obtaining real-time data that truly reflects community-wide usage of drugs—while concurrently assuring the inviolable confidentiality of every individual," he wrote. In 2005, a research team in Italy took up the challenge, using sewage to determine that official figures underestimated local cocaine use in the Po River basin, downstream from northern Italian cities such as Milan and Turin. Instead of the estimated fifteen thousand "cocaine use events" per month, they found evidence of about forty thousand per *day*, which the researchers called "staggering" in its economic impact.

The Italian project sparked a European research collaboration that now regularly tests wastewater throughout the continent and has advanced the science of wastewater analytics. Every illicit drug—and anything at all, really—we take into our bodies comes back out of us again. But researchers must identify the right chemicals, or biomarkers, for each drug they want to track. A good biomarker must come out of the body in a reasonably high quantity, not break down quickly in wastewater, and not come from other sources (such as animals). Metabolites—the chemical products that result from the body processing, or metabolizing, the drug—make the best biomarkers since they can't be confused with unconsumed drugs that have been dumped or flushed. The metabolite for cocaine is benzoylecgonine, which has no other source. Even with the right biomarker, the analysis isn't easy, as concentrations of illicit drugs in wastewater are about a thousand times lower than in urine, blood, or other body fluids. Researchers also have to take other factors, such as human mobility, into account: the European collaboration prefers to collect in March, when populations are more stable because people are less likely to take vacations and there are fewer tourists. Among other insights, its work has revealed that, in European cities, illicit drugs are just as common as licit ones. There's a lot of self-medicating going on.

Applied to drugs of abuse, sewage-based epidemiology, or surveillance, can provide useful data to help answer important policy questions, such as the effect of needle-exchange programs, "stop and frisk"-type policies, or legalization. With more targeted sampling from sewers, it could even help track down illicit activities like drug processing. But the science is replete with uncertainty and should be used in coordination with other epidemiological tools, such as surveys. And it's important, experts emphasize, to maintain high ethical standards. A big plus of this research is the anonymity of sewage at treatment plants, which prevents vulnerable individuals and communities from becoming targets of stigma. It's easy, though, to imagine an unscrupulous application that attempts to use sewage to link drug use or other activities to individuals, especially since there are currently no laws about how information from the sewer can be used.

Biobot Analytics' founders, Mexican computational biologist Mariana Matus and Iranian-Canadian architect Newsha Ghaeli, met at the Massachusetts Institute of Technology (MIT). They started by focusing their efforts on one of the United States' worst public health crises: opioid addiction, which is a leading cause of accidental death for Americans under fifty years old. Currently, public health officials get data on the crisis either from surveys, which underestimate use because people don't always tell the truth, or from overdose data, which are the tiny measurable tip of a larger, hidden scourge, since it's estimated that for every person who gets medical care for an overdose, there are possibly hundreds more afflicted with opioid use disorder. Making things worse, overdose data usually aren't available to officials until many months later. Instead, by looking at the markers of drugs in the sewage, Biobot Analytics can get a quick and unbiased measure of the use of prescription opioids such as oxycodone and codeine, illegal ones such as heroin and fentanyl, the overdose-reversal drug naloxone, and treatment drugs such as methadone and buprenorphine. Most of the markers are from urine, so that they can be sure that they reflect actual drug use, and they also measure nicotine, acetaminophen, and caffeine as points of comparison.

In 2018, Biobot ran a pilot project in Cary, North Carolina, after the town found the company online and won a grant. The mayor made fighting opioids a priority after the fire department responded to five opioid overdoses, three of them fatal, over the Thanksgiving holiday in 2016—a shocking tragedy for a wealthy town of 162,000 that claims to be one of the top ten safest communities in the country, going by crime statistics. Maps of the sewer system, matched to land-use and demographic data, helped the company and town officials identify ten manholes to target. Each manhole represented between four thousand and fifteen thousand residents—nearly 45 percent of the population when added together—and excluded buildings such as hospitals and factories, whose discharge into the sewers could skew the results. The sewage flowing by could be no more than about four hours old, so that the opioid metabolites wouldn't have time to break down. Cary employees lowered Biobot's box into those manholes for a day at a time in the summer and fall of 2018. Each day it was deployed, the bot processed about 2.5 gallons of the 200,000 to 800,000 gallons of wastewater passing by it.

Maps of overdoses had shown a hot spot in one part of the city. But the data Mr. Box collected revealed that opioid use was happening at similar levels throughout the town. To officials, that indicated that they should design programs with broader reach. The data also showed that different neighborhoods used different drugs—some illicit and some prescription—which helped officials tailor programs for each part of the community more narrowly. Levels of naloxone, the overdose-reversal drug, were higher than would be predicted from known overdoses, but also in roughly the same places as those known overdoses, so existing programs were likely getting the drug into the right hands.

As a result of the increased public outreach around the Biobot pilot, in 2018 the town saw the volume of unused prescription medicine brought to drop-off points roughly triple in comparison to the two previous years. That's a success, but the town is still working on how to act on the sewer data. "We took a 'Field of Dreams' approach: if we sample, actionable correlations will be found," writes Donald Smith, the town's wastewater collection system manager. "This was not the case." One possible approach, Biobot's research suggests, is to use the data to coordinate with pharmacies: to design take-back programs; to flag locations that seem to be dispensing more opioids than usual, perhaps to opioid "shoppers"; and to identify locations that need more training in dispensing naloxone.

While illegal drugs have been the main use of sewage-based epidemiology, researchers are hopeful about covering many more aspects of health, from smoking and drinking habits to nutrition to oxidative stress, a sort of stand-in for poor health because it's related to major causes of death including heart disease, stroke, respiratory infection, and lung disease. Scientists have already documented some forty-five hundred metabolites, connected to some six hundred conditions from obesity to cancer and infectious disease, in urine—and many of these could be targets for sewage epidemiology. Environmental engineer Rolf Halden of Arizona State University has coined the term *urban metabolism metrology* to describe his version of this vision.

Halden's Human Health Observatory collects and stores sludge samples—which are the solid products of wastewater treatment plants and not the raw sewage that flows into them—from more than 350 plants in the United States, which serve more than 40 million people, as well as another 150 plants abroad. With this resource, Halden has made his name, not only by looking for drugs, but also by looking for hazardous chemicals that may abuse us without our knowledge. These don't tend to enter our wastewater systems through our pee and poop, but through our household drains, since they're in the products we use, from cleaning liquids to nonstick treatments on pans to waterproofing layers on clothes. They also run off fields and streets into gutters. Wastewater treatment plants—designed for human waste—can't break down these hardy molecules, which flow out with the effluent or get embedded in the solids. Some of these are "very, very persistent," as Halden has explained, "so when we are long gone these chemicals will still be there." Frighteningly, the mix of these chemicals could be more harmful than any individual one on its own, contributing to what's known as the "body burden," as they build up and linger in body fat and organs.

Halden's sewage sludge work played a role in the banning of the antimicrobial chemicals triclosan and triclocarban. In the 1920s, chemists learned how they could transform molecules by stripping them of hydrogens and replacing them with the elements that are known as halogens (which include fluorine and chlorine), creating chemicals with almost miraculous properties—but also sometimes dangerous side effects. Over the years, many, such as DDT, got found out and banned, but triclosan and triclocarban, introduced in the 1960s for use in health-care settings, grew widespread in antimicrobial hand soaps and other products ranging from mouthwash to fabrics and even baby pacifiers. Although manufacturers convinced consumers that these were better than plain soap, that was not so—the antimicrobial action of the soap wasn't helpful to the average person, and most people didn't even wash their hands for long enough for the chemicals to do their work.

Because of their similarity to DDT and other substances like it, in 2002 Halden decided that he wanted to know more about the fate of triclosan and triclocarban. Developing new techniques to detect them, he followed them into the sewers of over 160 American cities and found that they didn't respond well to the processes in wastewater treatment plants that caused other compounds to degrade. Up to three-quarters of the chemicals got trapped in sewage sludge, some of which then got spread onto farmland. Then he looked in sediments downstream of the plants and also found the chemicals there. Near JFK Airport, in New York City, he found antimicrobials that had been deposited there when John F. Kennedy himself was president. Ultimately, scientists found them in breast milk and the urine of three-quarters of Americans. In animal studies, they disrupted the development of the reproductive system and metabolism—and many inferred that they could do similar damage in humans. The chemicals have the potential to push free-floating pathogens in the environment to evolve, leaving them resistant not only to the hand soap chemicals but also potentially to antibiotics. In 2014, Halden published a paper calling for their removal from consumer products. In 2016, more than a decade after he first outed the chemicals as widespread environmental pollutants, the Food and Drug Administration banned them.

The ban was a win, but Halden is haunted by the presence of so many almost indestructible chemicals whose effects on human health remain unknown and largely unexplored. Looking at the sewage sludge in his repository could be an alternative to testing these new chemicals one by one. That's because sewage treatment is roughly analogous to human digestion, so if a sewage treatment plant can't degrade a chemical, the body probably can't, either. And he has evidence for that: In one study, his lab compared more than a hundred chemicals that accumulate in sewage sludge to those found in the human body by the U.S. Centers for Disease Control and Prevention (CDC). As it turned out, about 70 percent of the chemicals in the sludge were also in humans—a surprisingly large number. "It tells us something that we should have known intuitively," he has said: "If a chemical makes it through a wastewater treatment plant and is not being degraded it also likely makes it into our body and is not being degraded."

Halden would prefer for the world to embrace a "green chemistry" that would not make hazardous chemicals at all. "If we have the means to make safer chemicals, why wouldn't we? We constantly improve the safety of the automobiles we make but fail to do the same for the chemicals we surround ourselves with and that we know will enter our body," he tells me. But, for now, sludge analyses could serve as a cheap alternative to running long-term health studies in humans, allowing regulators quick insights into new chemicals as they come onto the market.

Wastewater could also provide an early warning system for disease outbreaks. It has already made contributions to the fight against polio-myelitis, or polio, which transmits through fecal particles—so the poliovirus, if present, tends to end up in sewage. Eradicated from most of the world through an ambitious vaccination campaign, polio persists in a few remaining hot spots—parts of Pakistan and Afghanistan, with Nigeria declared polio-free in 2020—which tend to suffer from poor sanitation and where populations have resisted vaccination. Public health experts monitor outbreaks by counting cases of Acute Flaccid Paralysis—a sudden muscle weakness in the legs or other parts—but that syndrome occurs in just one of two hundred cases, which means that the metric misses many milder cases, allowing the disease to spread silently. It's a game of cat-and-mouse: as humanity gets closer and closer to eradicating polio, the virus becomes harder and harder to find with that metric.

Sewage, however, collected from sewers and other channels where feces flow, allows officials to see the silent cases. In 2013, a strain of polio made its way from Pakistan to some predominantly Bedouin communities in Israel. Israel's health department has had a sewage surveillance program in place since 1989, in which the country's Central Virology Laboratory analyzes sewage samples collected weekly from trunk lines and treatment plants. The lab caught the outbreak, and the health ministry contained it by administering oral vaccines. Similar programs currently run in India, Jordan, Lebanon, Nigeria, and Pakistan. Lab technicians add concentrated sewage samples to some cells that are susceptible to the virus, wait for a day, and then look in a microscope to see if they're affected.

Sewage could also help counter common pathogens such as influenza, even if they don't transmit through the fecal-oral route. If infected people pass traces of the pathogen in their poop, researchers can check for genetic residues in sewage. Alternatively, they can identify compounds to use as clever stand-ins. People with the flu tend to take over-the-counter medicines. Some government programs have tried to catch the early stages of a seasonal flu wave by monitoring sales of these medicines, but that's problematic, since people might buy the medicine in anticipation of flu season instead of in response to an actual infection. On the other hand, a metabolite of the flu medicine in the sewers would mean that someone popped the pill or drank the syrup.

Until recently, the field of sewage epidemiology and surveillance operated in relative obscurity, but that changed in March 2020, when the COVID-19 pandemic railroaded humanity. Early reports indicated that at least some infected people shed the virus in their stool. Scientists around the world began to develop methods for detecting the genetic material in wastewater and turning that into a tool for gauging rates of infection. The concept was particularly appealing since many countries—both high- and low-income—lacked the capacity to do enough direct individual testing to determine the scope and dynamics of the crisis. In the United States, the Biobot team seized the moment, turning their operations to focus on the new crisis. Publishing a "public call to action," they solicited wastewater samples from around the country. It was already too late to stop the first wave, but, just as they had tried with opioids, they hoped that their work could help evaluate interventions as well as catch new surges of infection early.

Data Stream

If Kashyap, the entrepreneur who ate whipworm, has his way, your toilet will also become a bot—though, instead of serving public health, it will serve you personally. The idea of a medical toilet first gained traction in the heady early 2000s, when the more-than-150-year-old UK brand Twyford announced plans to develop a toilet, called the VIP, that would contact a user's doctor at signs of trouble and also do the shopping. "If, for example, a person is short on roughage one day, an order of beans or pulses will be sent from the VIP to the supermarket and delivered that same day," a spokesperson said at the time. In the mid-2000s, Japanese toilet-manufacturing giant Toto released the Intelligence Toilet for the domestic market. With the help of a special bracelet, it measured the user's weight, blood pressure, body fat, and urine sugar, sending the data by Wi-Fi. It could also test the urine temperature, which would be handy for women charting their menstrual cycles. But the VIP never materialized, and the Intelligence is off the market—victims, Kashyap thinks, of unreasonably high prices (about ten thousand dollars per installation) and a failure of technology to meet the hype.

Now there's a new generation of medical toilets on the way. Research teams around the world are taking a variety of approaches, including for low-income contexts. And even Google seems to see a future in it, filing a patent that describes a toilet seat that can monitor heart health. (Is the Loo-gle in our future?)

In 2010, Kashyap, working alongside engineers, began trying to invent the toilet he wished he had; he founded his company, Toi Labs, in 2015. Since there are a lot of diagnostic tests of urine and stool, he started out by attempting to miniaturize them and somehow slot them into a toilet. It's "the first thing that people think of when they start to delve into this problem seriously," he says. But that proved difficult, and he wanted to develop something that could get to market soon. So he switched gears: instead of testing samples of pee and poop, the toilet would take data-rich images of them. "We're generating time-lapsed images that are in various spectra—in wavelengths—and we're interested in things like the color, the shape, the consistency, the clarity, the volume, the duration," he says.

The tests of urine and stool samples might come in future iterations, he says. One of the most promising avenues, Kashyap says, is to test the gases that rise off pee and poop, which could warn of disruptions to the microbiome. So a future toilet might not only see what's in the bowl—it could sniff it, too.

Instead of a whole toilet, Kashyap's product, called the TrueLoo, is a toilet seat that fits on existing toilets, which keeps the cost low compared to those earlier models—a few hundred dollars plus a subscription. Partnering with David Samuel, the founder of Brondell, an American toilet seat bidet company, Kashyap decided early on to target it first to the "senior living" market—that is, nursing homes. The data that comes from the TrueLoo, Kashyap says, could ultimately help senior-living operators discover and prevent "the kinds of things that just kind of happen day in, day out in senior living": urinary tract infections, falls due to dehydration, bloody stool or urine that could indicate a more serious problem, and infectious diarrhea from pathogens like norovirus.

In 2019 and 2020, Toi Labs installed 150 TrueLoos in nine senior homes, collecting more than 35 million images of some 185,000 toilet events, in order to train a machine-learning algorithm to recognize problems, as well as to evaluate the product's performance. "Our hope here is that through this technology we're going to be able to have a major measurable impact on the number of emergency room visits, the number of hospitalizations, and ultimately the costs associated with utilization of the health-care system," Kashyap says. Jennifer Bayard, a nurse and former executive at Carlton Senior Living in California, oversaw TrueLoo's trials in that company's communities. Since leaving in 2020, she works as an independent advisor, including, in exchange for some stock options, for Toi Labs. She predicts that the seat will be particularly helpful for patients with dementia, who may be able to use the bathroom without assistance but can't always report to care providers when they have problems. "I'd like to think that the TrueLoo is a voice for people who cannot communicate," she says.

Beyond senior living, Kashyap foresees an even bigger market, starting with people like him who have bowel disease. Kashyap, who no longer swallows worm eggs for his ulcerative colitis, has had a TrueLoo in his home for over two years, and tracking his stool characteristics with it has helped him manage his health. By analyzing the collected data, he made the discovery that the only dairy product that he can tolerate is homemade yogurt. "You could theoretically keep a log" of your bowel movements yourself, he says, but it's easier with the TrueLoo. Ultimately, Kashyap thinks that lots of people will want one in their home, so that they can get individualized insights into the "black box" that is our gut. "I want to know, like, when I start eating gluten, einkorn wheat, I want to know how that's actually affecting me. When I take out red meat from my diet, I want to know how *that's* affecting me. When I eat more collard greens, I want to know how it's affecting me. Yes, yes, yes!" Then, growing sober, he continues: "As I get older, if I start showing signs of bladder cancer, pancreatic cancer, colon cancer, I want this device to tell me and direct me and screen me to get the right tests."

If the concept works, a smart toilet could contribute to a revolution in medicine called "precision health," according to the late Sanjiv Sam Gambhir, who was chair of radiology at the Stanford School of

Medicine and director of the Canary Center at Stanford for Cancer Early Detection when he died of cancer in 2020. He became an evangelist for the field after a personal tragedy: his son, Milan, died of brain cancer at age sixteen. To Gambhir, precision health meant using all kinds of technologies "to prevent disease and, when that isn't possible, intercept and treat it earlier," he wrote. For years, he, engineer Seung-min Park, and colleagues worked on what they call a Precision Health Toilet; just a few months before Gambhir's death, they published a report on a technology that could identify individual users of a toilet—by, implausibly, the shape of their anuses, which apparently varies from person to person much like a fingerprint. (Park has noted that artist Salvador Dalí observed this feature of anuses decades ago.) Gambhir wrote:

> In a potential "smart home," one could sleep on bed sheets that monitor cardiopulmonary function, look into a mirror that measures one's vital signs using radar, use a toothbrush that performs biochemical analysis of one's saliva, and then have a smart toilet automatically analyze one's urine and stool for markers of disease. During the daily commute, sensors in the car can monitor stress levels and drowsiness through driving behavior, discourage driving if alcohol is detected in the breath, or detect outdoor pollution levels. Patterns of smartphone use reflecting depression or anxiety, such as a decrease in texting frequency, can alert the user to proactively address mental health through integrated self-care tools.

Gambhir stressed the importance of a high-quality health portal that will gather the data, analyze it for trends, and send recommendations— say, to perform a particular cancer screening—to the user's health providers, instead of leaving it up to individuals or a multitude of companies to sort through so much confusing and invasive information.

Smart devices make particular sense in gastroenterology, where taboos get in the way of doctor-patient communication. Even where patients aren't reluctant, they often can't recall much about their recent bowel movements. So data on frequency, color, and consistency from a smart toilet might be more reliable than patients' responses. And patients couldn't forget to check in with their smart toilet every day, as they might with an app or device they have to wear. But as with all of these technologies, there are risks and downsides. Privacy is at stake when we entrust our personal data to companies. "One immediate privacy concern that we have is that people's genitalia may be imaged," Kashyap says. "We proactively generate a privacy zone where we essentially create a block on the area that may be imaged. And, in the event that it is imaged, we're able to expand that privacy zone."

But any device that gathers health data presents a risk if those data get out. When, in April 2018, Kim Jong-un became the first leader of North Korea since 1953 to cross into the demilitarized zone between North and South Korea to meet with South Korean president Moon Jae-in, and again in June 2018 when he met with Donald Trump for a nuclear disarmament summit in Singapore, he took his own toilet. Apparently, one goes with him everywhere, sometimes in a special toilet car, so he doesn't have to use shared facilities. "The leader's excretions contain information about his health status, so they can't be left behind," Lee Yun-keol, a former member of a North Korean Guard Command, told the *Washington Post*. If that sounds a bit paranoid, consider that Stalin is said to have used a special toilet to steal excretions from world leaders, including Chinese leader Mao Zedong, and the CIA and MI6 may have tried (and failed) to get a stool sample from Mikhail Gorbachev. Whether any useful information has ever come to light through this kind of spycraft, however, remains a secret.

Finally, there's the risks involved in all kinds of health screenings. One false positive (a screening test indicates disease where there isn't any) can send healthy people down a road of anxiety and even unnecessary treatment, and too many false positives can desensitize them to alerts that they should act upon. False negatives (a screening test fails to indicate disease where there is one) can lull sick people with a false sense of security. Even when working well, smart toilets are likely to cost more than standard toilets, which would potentially contribute to increasing inequality in health care.

Kashyap believes there will be a strong market for this product, at the right price and in the right place. At first, he thought that he needed to base his company in Japan, the home of the advanced toilet. Eventually, however, he realized that Americans would need a TrueLoo even more. Unlike in Japan, he says, "the American health-care system is not looking out for you proactively." So your toilet should.

Pipe Down

There's more than one way to transport poop.

The sewer is the conscience of the city.

—Victor Hugo,
Les Misérables (1862)

System Critical

My first night in Cap-Haïtien, Haiti's sewer-less northern port city, I have just returned to my hotel from dinner when I hear a roar like a jet engine. Alarmed, I run to the window. It's rain—torrents of it, arriving in thick waves to pummel the tin rooftops.

The next morning, I head out into a postdiluvian mess. The day's main destination is Fort St. Michel, a poor neighborhood of forty thousand near the city's airport. Like nearly half of the city, it is in a floodplain, and the narrow streets have become stagnant canals, with plastic bottles, tin cans, and rotting bananas floating in them. Families of small black pigs walk around, pissing into puddles. Barefoot children wade to school in their uniforms, holding their shoes so as not to dirty them. A gallant teenage boy carries a giggling girl on his back as the neighbors tease them.

Workers at the nonprofit SOIL in Haiti load
empty pails in a flatbed truck at a depot.

With me in the rental SUV is Sasha Kramer, executive director of
the nonprofit SOIL, which runs an unconventional sanitation business that serves this and other poor neighborhoods that otherwise
rely on pit latrines or the uncertain privacy of dark corners and the
spaces between cars. We're following a plucky green-and-red vehicle,
a three-wheeled motorcycle fitted with a flatbed on the back, which
is making the rounds. It's slow going: the motorcycle driver struggles
to find a way through, inching forward and then backtracking when
roads prove impassable. When he stops, his colleague, in a neon-green
safety uniform with thick rubber gloves, wades up to his thighs in the
waters, balancing stacks of tightly sealed green plastic pails in his arms.
He places empty pails at doorways and on corners and picks up full
ones left there for him. In those full pails? Poop.

Since Roman times, the ideal of sanitation infrastructure—indeed of all
urban infrastructure—has been sewers. For some, they have epitomized
civilization: Victorian thinker and philanthropist John Ruskin declared
that "a good sewer was a far nobler and far holier thing . . . than the

most admired Madonna ever printed." More recently, the website Atlas Obscura dubbed the sewer "the most marvelous everyday invention" in its Mundane Madness competition. (Incidentally, the final four included the can opener, paper, and, in a related entry, toilet paper.)

Yet the smelly, lawless tunnels have just as often represented the dark side of cities. The ancient Roman Suetonius wrote that the Emperor Nero stabbed men on their way home from dinner and dropped their bodies into the sewers. Another ancient claimed that the emperor Elagabalus was murdered in a latrine and his assassins tried, unsuccessfully, to stuff him into a small sewer opening. Later literature would also use them in similar ways: Victor Hugo, a sewer aficionado, has good guy Jean Valjean escape through grimy Parisian sewers in *Les Misérables*, and Orson Welles's bad guy Harry Lime does the same in the postwar Viennese sewers of *The Third Man*. More recently, the Teenage Mutant Ninja Turtles live in the sewer and the Super Mario Brothers (who are plumbers) have taken it as the setting of many battles with monsters. Sewers have been used for staging crimes, sheltering from air raids, and as command posts. "The history of men is reflected in the history of sewers," wrote Hugo. "Crime, intelligence, social protest, liberty of conscience, thought, theft, all that human laws persecute or have persecuted, is hidden in that hole."

The Romans began building sewers in the sixth century BCE, with the giant Cloaca Maxima (meaning "Great Sewer"), a wonder of nearly eleven-foot-high stone vaults. But this underground cathedral wasn't meant to transport waste; rather, its function was to drain the marsh on which the city of Rome was built. Over the centuries, the Romans expanded the system throughout the city center, draining more land—for the Colosseum and the Circus Maximus—and controlling dangerous stormwater, as well as handling huge volumes of wastewater from installations like the Baths of Agrippa and the Pantheon, some of which would have included latrines. Today, many sections still transport wastewater and you can see the triple-arched outfall of the Cloaca Maxima along the

Tiber River. The famous Bocca della Verità, a stone disk decorated with a face, was very likely originally an ancient manhole cover. Newlyweds stand in long lines to stick their hands in the open mouth as they pledge to be faithful to each other, since it's said that it will bite anyone who lies to it.

Koloski-Ostrow, the classical archaeologist, has investigated how well these sewers worked and found them lacking. For one, filth built up in the channels and still does, as she's seen for herself, writing:

> Once I tried to enter the sewer behind the Church of Santa Maria in Cosmedin and found the silt completely blocking the entrance. In another expedition in a sewer channel under the Roman forum (about 2007), I found myself repeatedly tripping over piles of debris and silt as I tried to maneuver my heavy boots along the slippery ledge through muddy waters still draining into the sewer from modern streets.

The Romans didn't understand the physics behind their invention, so they couldn't prevent this buildup through better design of the tunnels or junctions. Instead, slaves and convicts probably cleaned them out.

There was another big difference between the Roman sewers and today's: although some public latrines would have drained into them, most people didn't connect their private toilets—perhaps because they didn't want to. Looking at it from the Romans' perspective, this makes sense. Toilets didn't have any type of seal, so any connection to the sewer meant that substances could come up as well as go down. Among those: stench, floodwaters, and critters like rodents and insects. Decomposing organic matter in the sewers would have generated hydrogen sulfide and methane, and Koloski-Ostrow claims that explosions could have singed behinds. One ancient writer described a monstrous octopus that entered a merchant's home through the toilet and ate pickled fish from his pantry.

It took almost fifteen hundred years after the Roman era ended before a London lawyer named Edwin Chadwick helped create a vision for sewers as technologies especially for human waste, transforming the city's underworld into what Hugo would describe (referring to Paris, which took a parallel path) as "clean, cold, straight, correct. . . . It resembles a tradesman who has become a councillor of state." In the early nineteenth century, sanitation was largely left up to individuals, and outhouses, often known as privies, prevailed. Then, as piped water connections increased and more people adopted water closets, wastewater began to overwhelm cesspools and storm drains.

Born in 1800, Chadwick believed in the environmental roots of disease, though people didn't yet understand the mechanics of disease transmission at this time. To improve the environment of the city, Chadwick thought it would be necessary to replace the patchwork of water pipes and privies with a connected, comprehensive hydraulic system that would bring clean water into homes on one hand and then take dirty water out to sewers and ultimately to agricultural lands on the other. This blueprint for an "arterial-venous" city would, as Chadwick wrote, "complete the circle, and realize the Egyptian type of eternity by bringing as it were the serpent's tail into the serpent's mouth."

Building such a system would be an engineering feat of enormous proportions. One challenge then, as now, was to design the sewer pipes to get an optimal flow of water. And so velocity became Chadwick's "chief obsession," Martin Melosi writes in *The Sanitary City*. On the one hand, the water must flow quickly enough to prevent suspended particles from settling at the bottom, which clogs the sewer. On the other hand, if the water flows too quickly, the suspended particles scour the interior of the sewer, potentially damaging it; a powerful flow can even blow out a weak structure. At first, Chadwick advocated shaping brick sewers like downward-pointing eggs. Later, he changed his allegiance to glazed earthenware pipes. But his plans tended toward the idealized and ignored on-the-ground realities of heavy rainfall, improper use (like flushed cloth and paper), and poor installation.

The British Parliament never bought into Chadwick's all-

encompassing, integrated, circular water and wastewater system, but
it did fund a massive sewer network in the 1860s, after London experienced the Great Stink from the Thames. To build it they tapped
Chadwick's rival, engineer Joseph William Bazalgette, who disagreed
with many of Chadwick's preferences. Instead of transporting wastewater to agricultural lands, Bazalgette continued to dump sewage in
the Thames, though he employed "intercepting sewers" to move these
outfalls downriver. And he wasn't a pipe purist: he used a combination of brick and earthenware. In the end, the groundbreaking network
contained eighty-three miles of sewers, plus a few pumping stations to
overcome gravity and tides, and served as a model for the world. Despite
the failure of his most ambitious ideas, Chadwick could take a good deal
of credit for this outcome, Melosi writes. "One of the key results was
the transformation of Victorian cities from their Dickensian bleakness
into more livable environments."

An 1883 portrait of engineer Sir Joseph William Bazalgette, who designed
London's sewer system, originally published in *Punch* magazine.

*   *   *

Other rapidly growing cities adopted the London model. In America, perhaps the most remarkable sewage works took place in Chicago. The city faced urgent problems: in 1854, a cholera epidemic felled roughly fourteen hundred, or one in every eighteen, residents. Since the terrain was so flat, engineer Ellis S. Chesbrough suggested raising the city by as much as twelve feet in some areas, so that sewers could work by gravity. In 1858, some two hundred jackscrews hoisted a giant brick building by more than six feet. In the following years, engineers raised a variety of other structures, including a luxury hotel, an ornate iron building, and half of a city block, with the stores on it still open during the procedure. Engineers elevating the streets laid sewer pipes, covered them with dirt, and then paved over them. These connected to a large intercepting sewer, which transported the wastewater to the Chicago River, which was dredged to make room for the coming load.

Without any wastewater treatment, the sewage flowing into the river and out to Lake Michigan soon threatened the city's drinking water. Ultimately, the city undertook another major engineering project, reversing the flow of the Chicago River so that, instead of discharging into Lake Michigan, it drew in water from the Great Lake and discharged it west into the Mississippi River watershed. This diversion, however, just passed Chicago's filth to other cities downstream until it added wastewater treatment.

At first, sewers spread because cities would take on debt to pay for them, thanks to new financing mechanisms. In England and Wales, local authorities accumulated almost 100 million pounds of debt for waterworks and sewers by 1905—adjusting for inflation, that's equivalent to about $16 billion today. Between 1860 and 1922, Melosi writes, American municipalities grew their debt from $200 million to more than $3 billion. This investment has been worth it in lives saved and public health costs averted, though new investments are now needed to maintain and sometimes replace these aging systems.

(A sewer pipe's life span is about fifty to a hundred years, depending on the type of material and the conditions.) And sewers never became truly universal in the United States. After World War II, a building boom created suburban homes for the millions of GIs returning home. But these came without water and sewer connections, so people installed wells and septic tanks—technologies more appropriate for rural farm life, Melosi writes. In the countryside, "a little odor or a soft spot in the middle of a field far from habitation was no cause for concern," but "septic tanks when brought to town were a very poor choice on these small 'postage stamp sized' lots." Many suburbanites now face the high cost of upgrading failing septic tanks, which contaminate groundwater with microbes and nutrients. Discharge from sewage and septic systems still contributes a substantial amount to nutrient pollution, which the U.S. Environmental Protection Agency (EPA) calls "one of America's most widespread, costly and challenging environmental problems." In some places, such as northern Florida and Cape Cod, it's the main source of that pollution.

This gap has replicated itself on a global scale, as sanitary imperialism exported the sewer model ideal but not, among other things, the necessary financial mechanisms that allow cities to borrow the high up-front costs. Today, only about 62 percent of urban dwellers worldwide have access to sewers—a percentage that remained essentially flat between 2000 and 2017—and those people are mostly in high- and upper-middle-income countries, where coverage is still increasing. In low-income countries, the sewer coverage dropped from about 24 percent to about 17 percent in the same period due to urban population growth. Many cities are so dense and growing so quickly that it's difficult to imagine undertaking the works necessary to install conventional underground sewers, which is why there's such an urgent need for other options. The world has been slow to admit it, but the sewer boom has ended, and it did so long ago.

Redesigning from the Bottom Up

In the classic 1965 science fiction novel *Dune* by Frank Herbert, the inhabitants of a bone-dry planet wear "stillsuits," garments that reclaim all moisture from body waste, recirculating water from them back to the wearer to drink. The suits process urine and feces in the thigh pads, as one character explains: "With a Fremen suit in good working order, you won't lose more than a thimbleful of moisture a day." This, perhaps, is the most extreme vision of what's often called distributed, or decentralized, wastewater treatment, an idea that's gaining traction in both low- and high-resource contexts alike.

More realistic—but still futuristic—are some of the toilet concepts funded by the Gates Foundation. No person has come to embody and influence the new toilet revolution more than the visionary sixty-something clean-shaven, preternaturally optimistic, ultrawealthy Gates. He started down the toilet path around 2008, after he stopped working full-time at Microsoft and started playing a larger role in the Gates Foundation, now among the largest private foundations in the world. In their travels, he has explained in public statements, he and Melinda, his wife, saw the realities of poor sanitation for themselves. After reading an article by *New York Times* journalist Nicholas Kristof, the couple realized that this basic need had to be met before their foundation could hope to achieve its other goals for human health and well-being.

Gates didn't see a future in large infrastructure like sewers and wastewater treatment plants, however. They are too expensive to build and operate and require resources like water and electricity that many places don't have in abundance. (And, as others have observed, the "helicopter" approach to development, in which funding agencies drop in, build big conventional infrastructure, and get out, all within a few years, has proven disastrous. When the local government fails to find the money to maintain the infrastructure, it breaks, never to be repaired. These white elephants litter many lower-income regions.)

Gates thought about what he had done for computers—transform giant, expensive mainframes into personal devices—and wondered if he could do the same for toilets, not by inventing himself but by funding projects. "In 2009, I posed a question to a group of scientists and engineers: Was it possible to leapfrog the long-accepted 'gold standard' of sanitation—flush toilets, sewers, and treatment plants?" he explained at the Reinvented Toilet Expo in 2018 in Beijing, China. The term *leapfrog*, now almost a cliché, was bracing at the time: it suggested that poor people did not need to follow the path of Western sanitation, from open defecation through pit latrines to sewer-based sanitation, but could instead jump to something better, something that might even make the rest of the world envious. He often made an analogy to the mobile phone, which had allowed people in poor countries to get telephones without ever having landlines.

But how to turn that idea into a reality? Gates focused on the technology itself, pouring $200 million in seven years into making the leapfrog happen, and pledging $200 million more. (To the foundation, that's not a whole lot: in 2018 it spent about $5 billion on grantees across all of its funding areas.) Its most well-known sanitation initiative is the Reinvent the Toilet Challenge, first announced in 2011, which provided research and development funding for university-based teams. The competition required that the off-grid toilets treat the waste on-site, at a low cost, and without piped-in water. The winning team from the first round, announced in 2012, came out of engineer Michael R. Hoffman's lab at the California Institute of Technology. Instead of relying on microbes, as conventional treatments do, it used electricity to clean the wastewater. Electrochemical disinfection works by passing a current between two submerged electrodes—sort of like the common science-class experiment. This splits molecules, creating new chemicals, including chlorine, that turn the liquid into a bleach-like solution that kills off pathogens.

Although the challenge fascinated the media and the public, in the development sector, Gates's arrival on the toilet scene created a bit of consternation. Some appreciated the attention that it brought to the problem, but most thought that what Gates was asking for was absurd. "People said, they're crazy, you know? They want a toilet for outer space to be in the slum for only five cents? How can that work?" designer Harald Gründl of the firm EOOS, who collaborated on a project that won an award in that first competition, told me at the time. The implication that the people already working on sanitation weren't somehow smart enough offended some of them, when the reality was that they were wildly under-resourced. Still others worried that the mogul's technophilic, corporate vision—not to mention his larger-than-life personality, intellect, and checkbook—would crowd out more grounded insights and solutions. Today, most of the reinvented toilets remain easier to envision in Gates's futuristic house, which he calls Xanadu 2.0, than in slums or remote countryside.

But Gates has persisted, expanding the funding programs beyond the original challenge to technologies and programs better suited for community toilets or whole municipalities, including those that better manage the sludge from existing pit latrines and septic tanks. The foundation's strategy also extends to developing new service models for sanitation in cities, as well as improving government policies and regulations and training a new generation of local engineers and practitioners. "It's no longer a question of if we can do it," Gates said in China. "It's a question of how quickly this new category of off-grid solutions will scale."

In some places, unconventional sewer grids can transport sewage better or more cheaply than conventional sewers, often connecting smaller numbers of households to smaller-scale treatment plants. For arid or cold places, vacuum sewers work well, because they require little or no water. On a parched, palm tree–shaped artificial archipelago called Palm Jumeirah in Dubai, more than two thousand luxurious villas have been connected to twenty-five miles of vacuum sewer line, providing wastewater services to the people who live on one of the world's largest artificial islands. It depends on negative air pressure generated at a central facility to suck, instead of push, the toilet waste through the system. This idea is not new—a Dutch engineer first tried it out in the 1880s—but recent advances have made vacuum systems more reliable and cheaper, and they now serve communities as diverse as a seven-hundred-resident native village of Savoonga, on an island in western Alaska, and a marshy, ninety-five-hundred-resident municipality near the capital of Slovenia. In Dubai, the clean water that comes out of the advanced system gets reused for landscape irrigation.

For hilly places, pressure sewers work with pumps, sending wastewater uphill, which is great for waterfront properties, since gravity sewers and septic systems would drain downward toward lakes and rivers. In each household, waste flows to a collection tank, where a pump grinds it up into a slurry and shoots it into the pipes, which can be narrow and shallowly buried, meaning that installation costs about half that of a gravity sewer. But both vacuum and pressure sewer systems require a constant supply of energy to run and expertise to maintain, making them impractical in places. In 2009, a vacuum sewer seemed like a good idea for an informal settlement called Kosovo in Cape Town, South Africa, because of the flat, sandy terrain and high groundwater levels. The city installed 354 communal toilets in 43 blocks, but people misused them—flushing utensils, bricks, and other stuff—and technicians didn't know how to fix them, leading to equipment filled with reeking sewage.

Victorian engineers thought that gravity sewers had to be large and at a steep gradient, but more recent advances have shown that very narrow, shallowly laid pipes can provide a lot of the advantages of larger ones at lower costs. In the 1980s these caught on in Brazil, where they're called condominial sewers. Instead of providing connections to each house, a condominial system provides one connection per city block (known as a condominium). The neighbors in the block then organize the best way to hook up their houses to that connection. They also sometimes take part in the building and maintaining of that private line. This turns the city block into a social unit that takes collective action. In Brasília, the capital, condominial sewers serve five hundred thousand people, ranging from the very poor to the very wealthy. In some cities in Brazil and elsewhere, however, attempts at similar programs have failed, largely because of problems managing and paying for the system.

Solids-free sewers transport just the liquid fraction, leaving the solids in storage tanks at the site of the toilet. In Nala, Nepal, a town about an hour's drive from Kathmandu, residents did much of the work for this type of system themselves, digging shallow trenches and laying the narrow pipes for a solids-free sewer for twenty-three hundred people. To treat the water, they installed a sort of advanced septic tank, followed by a constructed wetland, which mimics the water-cleaning abilities of natural wetlands. In similar communities around the world, innovative small-scale treatment systems dry solids in beds planted with large grass species that can, once harvested, serve as fodder for livestock; farm fish in ponds that treat effluent; or even bake sludge into bricks.

Even some major U.S. cities with centralized infrastructure, such as San Francisco; New York; Austin, Texas; and Portland, Oregon, have started allowing and even encouraging decentralized waste-water treatment systems, often on rooftops or in basements, which recycle water and generate gas or fertilizer. Although the utilities give up some revenue, they gain flexibility, since large systems can't nimbly expand or contract to accommodate changes in population and demand. Another advantage is resiliency: During Superstorm Sandy in 2012, storm surges knocked out many low-lying facilities in New York and New Jersey—one Long Island plant was offline for a month, during which hundreds of millions of gallons of raw sewage spilled into waterways. In contrast, dozens of distributed wastewater systems in the region managed to get back online within twenty-four hours with the help of generators.

In the United States, there are over eight hundred thousand miles of public sewers, as well as five hundred thousand miles of private sewers that connect private property to public sewer lines, according to the American Society of Civil Engineers' Infrastructure Report Card. Wastewater treatment systems will need to accommodate some

56 million new users by 2032, at a cost of at least $271 billion. Investing some of that in innovative distributed projects might be the way to raise the nation's poor wastewater grades.

The Dirtiest Jobs

In February 2020, news of a death in Delhi, India, reached the press. A private contractor, working for the Delhi Development Authority, had engaged five workers to unblock an overflowing sewer in one of the city's oldest districts. One went down fifteen feet without safety gear, armed with only a rope and a wooden stick. When he stopped responding, another followed. When the second didn't come out, the others called the police, who pulled the men out with ropes. The first, twenty-four-year-old Ravi, died of suffocation by toxic gases, and the second, thirty-five-year-old Sanjay, was rushed to the hospital, unconscious but alive. Tragic, avoidable, yet few were surprised by this news: the year before, at least 110 sewer and septic tank cleaners in India had died from accidents while doing this dangerous job.

So long as there have been sewers and septic tanks, someone has needed to clean them—and that someone has almost always been at the very bottom of the social ladder. In Roman times, Koloski-Ostrow writes, "we can speculate that the salary paid to a *cloacarius* (trained sewer-worker) could not have provided for more than a meager daily subsistence, perhaps twenty-five denarii per day. Such men were subjected to nightmarish pathogens in the wastewaters of the sewers." In medieval Europe, "gong farmers" emptied pit latrines and often employed boys to enter small spaces. Still today, in many parts of the world, informal workers who clean and empty sewers, septic tanks, and toilets live with intense stigma and carry nicknames such as frog-men, sweepers, and scavengers. According to a report by the World Bank, the World Health Organization, and other organizations, these workers suffer from a litany of health problems, including "headaches, dizziness, fever, fatigue, asthma, gastroenteritis, cholera, typhoid, hepatitis, polio, cryptosporidiosis, schistosomiasis, eye and skin burn and other skin irritation, musculoskeletal disorders (including back pain), puncture wounds and cuts, blunt force trauma, and fatality."

In wealthy countries, safety regulations and mechanization have blunted most of the hazards of those jobs and sewer workers tend to be public servants with decent pay. Many in the development sector seem to have assumed that poor countries would follow this same trajectory as they pushed for universal sanitation. The priority was getting toilets to users and, later, making sure that the sludge from those toilets got to treatment. They didn't suppose that the spread of safer sanitation would also lead to the proliferation of dangerous, degrading work.

In India, ancient rules delegate the handling of human waste to a particular subcaste of Dalits (once known by the derogatory term *untouchables*). Traditionally, Indians would defecate in a spot or structure in or around the home known as a "dry latrine." They would then hire the Dalits to collect the waste in a basket and dispose of it outside the village—a practice known as manual scavenging. Traditional manual scavenging is now illegal but still continues, and the Dalits assigned to do it must also make their livings cleaning septic tanks and sewers, often without protective equipment or machinery. (Typically, women do the dry-latrine cleaning, while men clean the septic tanks and sewers.) Everyone else just looks away, says social scientist CS Sharada Prasad. "I was raised as someone who wouldn't even see these things, though sanitation work is happening all around me."

Today, members of India's manual-scavenging caste have become collateral damage of what, in many ways, is the biggest success story in sanitation in the past decades. Building on the insight of Mahatma Gandhi that "sanitation is more important than political independence," Prime Minister Narendra Modi's high-profile Swachh Bharat, or Clean India, Mission aimed to completely end open defecation in the country, which has long counted the largest number of people who don't use toilets in the world (though, because the country is so populous, not the highest percentage). Costing billions of dollars and running from

2014 to 2019, it provided subsidies and employed behavior-change programs to spur the building of new latrines for hundreds of millions of people, particularly in rural areas. Major Bollywood stars took part, making movies glorifying toilets and menstrual pads.

In addition to praise, however, the Swachh Bharat Mission also generated a lot of scrutiny and anger. Modi exaggerated the results, declaring the country "open-defecation-free" in 2019 when it was obviously not, some open-defecation-free zones have seen backsliding, and the toilets were often poorly made, leading to pollution. In an attempt to shame people into abandoning open defecation, some officials and others photographed women and assaulted men who were breaking the rules. At the same time, Modi pursued a sometimes repressive and violent ethnonationalist agenda that used the concept of "cleanliness" against the country's minority populations, who were smeared as dirty. "Clean India" did not only mean more toilets; it also sometimes seemed to mean fewer Muslims.

When, in September 2019, the Gates Foundation bestowed upon Modi its Global Goalkeeper Award for fulfilling his commitment to sanitation, the blowback was fierce: a hundred thousand people signed a petition requesting that the foundation withdraw the honor. Actors Riz Ahmed and Jameela Jamil, who were set to attend the event, pulled out. High-profile sanitation experts objected. "Modi's sanitation campaign has no doubt benefited people, but how can access to a clean toilet outweigh the violence and persecution they may face in the rest of their lives?" lawyers Suchitra Vijayan of the Polis Project and Arjun Singh Sethi of Georgetown University Law Center wrote in the *Washington Post*.

Another problem is that the Swachh Bharat Mission gave little attention to the people who do the real work and bear the harshest burdens of keeping India clean. The recent rise in sanitation-worker deaths in India might be an artifact of better reporting due to increased public attention to the problem, but many worry that it is the direct result of all the new toilets.

*   *   *

If history is a guide, nobody is going to hand rights to sanitation workers without a fight. In 1968, more than a thousand of Memphis, Tennessee's sanitation, sewer, and drainage workers went on strike. The immediate trigger for the strike was the death of two garbage collectors who were crushed in a truck's compactor on February 1. Around the same time, a smaller, everyday tragedy had occurred: twenty-one sewer workers in the Public Works Department had been sent home without pay due to heavy rains, while their supervisors stayed and got paid—a common event that strained worker finances and highlighted racial disparities, since the workers were Black while the supervisors were white. Calling for union recognition, the workers began a strike on February 12; famously, workers carried signs that read: "I AM A MAN." In April, civil rights leader Martin Luther King, Jr., went to Memphis to support the strikers, giving his famous "I've been to the mountaintop" address. Then the worst happened: just a few days later, as he stood on the balcony of Memphis's Lorraine Motel, a gunman shot him dead. When the city council finally agreed to recognize the union, triggering an end to the strike, the strikers had gone two months without paychecks and lost their hero and champion.

Today, more sanitation workers worldwide are finally getting a seat at the table due to both grassroots organizing and attention from groups such as WaterAid. I saw some evidence for myself at a 2019 sanitation conference, called, plainly, Faecal Sludge Management, in Cape Town, South Africa. Usually, conferences such as these are places for experts from universities, nonprofits, and government agencies to mingle. But the Cape Town conference invited teams of sanitation workers from around Africa to share their perspectives and to have some fun. Competing for glory in a pit-emptying challenge, the teams put on protective gear, shoveled artificial sludge—consisting of everything from sand to diapers to hair weaves—from plastic bins and transported it by handcarts to a low stage in the convention center, while the other attendees cheered them on. The winners came from the eThekwini Municipality in South Africa, which experts consider a case study for good sanitation labor practices: by law, employers there must give sanitation workers education and protective equipment.

In India, two related hurdles remain: the existence of dangerous and inhumane sanitation work, and the expectations that Dalits will do it. One partial solution to the former is to give workers machines. But Sharada, who has followed vacuum truck workers for his fieldwork, warns that technologies don't always help if they're not part of a larger focus on infrastructure, training, and workers' rights. Because of rampant poor construction, "a pit can be a solid monolith of sludge. And you just cannot empty it using a truck." Workers with vacuum trucks don't realize the hazards they face because of a phenomenon called mediation by mechanization. "They wash their pipes and everything else, [but] they don't wash their hands," he says.

More important, the occupation of sanitation worker should be decoupled from caste, argues activist Bezwada Wilson, who founded an advocacy organization called Safai Karmachari Andolan. Although he never worked as a manual scavenger, he has said, some people still call him bhangi (sweeper) because of the circumstances of his birth. The enduring stigma means that it's difficult for people of the caste to find any other work—and also that few members of other castes will do it. "It is never said that [manual scavengers] can engage in only one particular occupation. But it is very clear that if they leave that occupation, they would automatically lose their livelihood," according to Wilson. "The right to sanitation cannot be just about the rights of users. It must also include the rights of the service providers."

The New Night Soil

Among the most unlikely of today's toilet innovations are simple containers, like the plastic pails collected by SOIL, Kramer's program in Cap-Haïtien. There's a precedent for these systems in some nineteenth-century European cities, where "night soil" collectors would travel through the streets after dark to collect metal pails or wooden barrels of waste from homes and take them in wagons to the countryside as fertilizer for farmland. In the Netherlands, where night soil thrived, the last barrels went out of service as late as 1983. By then everyone—perhaps especially sanitation experts—viewed them as unsafe, undignified, and generally just icky. The bad past, not the bright future.

Kramer, however, holds a Ph.D. in ecology, not sanitation. An American born in California, raised in New York, and now living in Portland, Oregon, when she's not in Cap-Haïtien, she has an unstudied hippie vibe: big earrings, flip-flops, jean skirts. She named her young son Biko, after South African anti-apartheid activist Steve Biko. She speaks fluent Haitian Creole and, following the Haitian custom, greets everyone with a kiss on the cheek—even strangers—which, since she works with poop, seems something like a dare to me. After a 2004 coup ousted President Jean-Bertrand Aristide, whom she deeply admired, she headed to Cap-Haïtien as a human rights observer, meeting there a group of foreigners and Haitians who shared her interests. Two years later, when she finished her degree, she moved to the country with the goal of setting up composting toilets. To Kramer's thinking, treating human waste this way is "a really nice way to use ecological systems to address a whole suite of basic human rights issues." Not only health and safety but also low national food production that arises in part from Haiti's depleted soils.

Yet Kramer didn't start with poop pickups—that concept arose over time, out of necessity. At first, following examples from South Africa, she and her colleagues built public toilets perched on big composting vaults in rural areas. Despite initial enthusiasm, the communities didn't maintain the toilets. "People who are struggling just to survive just don't have time to clean up someone else's waste," Kramer says. In January 2010, a devastating 7.0-magnitude earthquake hit, leveling Port-au-Prince. Desperate to help, she and four colleagues took off for the capital. Kramer knew little about emergency response, but there were giant tent cities full of displaced people who needed toilets.

With funding from Oxfam Great Britain, she tried something new: instead of leaving the waste in the toilets to compost, her team stashed movable containers inside of them. At first, they used fifty-five-gallon drums, but these proved too heavy and awkward to move when full. So they settled on fifteen-gallon plastic pails. About a hundred people per day used each of SOIL's two hundred toilets, and Kramer's team collected the pails weekly. They bought a piece of land and set up a composting site. At the time, it was the only functioning human-waste treatment site in all of Haiti. Its first sale of compost helped transform a tent camp in Port-au-Prince back into a soccer field.

Then came cholera in the fall of 2010, spreading in the country for the first time in modern memory. Kramer's husband was one of the more than half a million people who came down with it; he recovered, but thousands died. At the start, the epidemic seemed just a second bout of terribly bad luck, but later it would come out that United Nations peacekeepers, sent to help with earthquake response, had inadvertently brought and spread it by using toilets that discharged directly into a river. It made SOIL's work all the more important, though they had to toughen up on safety during collection and processing. Before cholera, "we were, like, poop is cool. Poop is amazing," Kramer says, recalling how she would work elbow-deep in half-composted shit in her sandals. "We had to refine our message to say, poop is really dangerous. And the things you can do with it are really amazing."

A toilet at SOIL's headquarters in Haiti.

*   *   *

In 2011, Kramer decided it was time to bring the lessons from Port-au-Prince back to SOIL's original home in Cap-Haïtien, to see if a container-based toilet service could work in nonemergency situations. This time, the service would be for households, because people would be more willing to pay for and clean a private toilet in their own homes. The service would also provide jobs instead of relying on volunteerism. A team from Stanford University designed the first of the toilets in consultation with residents of a poor, dense neighborhood called Shada that was known for toilets that overhung the river, as well as the use of plastic bags known as "flying toilets," since people would dispose of them by flinging them through the air.

There's a small exhibit of design iterations in a gazebo on the grounds of SOIL's headquarters, a sparsely furnished villa outside town. Many of the early versions were simple wooden boxes, charmingly painted but doomed by problems such as maggot infestations or terrible odors. Ultimately, customers preferred "something that sort of resembles a porcelain water flush toilet," Kramer says. From the business side, Kramer needed something durable and cheap, and that they could hire local people to make.

The latest version sits not in the museum but in the headquarters' bathroom. You wouldn't mistake it for a porcelain toilet since it's made out of a cement-based plaster applied over metal mesh, which makes it very light. But it's white with a recognizable curve in front and plastic seat on top. Opening the lid, I see a green pail inside, but any poop in there is hidden in a fluffy, odorless mass of peanut shells and bagasse, the fibrous leftovers from processed sugarcane. This is what SOIL composts. That day, all I have to offer is pee, which SOIL doesn't collect, since transporting all that heavy liquid would be far too costly in terms of fuel. A funnel in the front diverts it away. With no water in the system, there's no tinkle and no flush—in the calm, private context of the headquarters, I would even say it's peaceful.

It's the next morning, after the rainstorm, that Kramer and I, along with a translator and photographer, follow the SOIL team into Fort St. Michel. We chat with some customers, many of whom have requests for the SOIL team. One man tells Kramer that he would like more empty containers, as well as a lock on his toilet to prevent others from coming to use it when he's not at home. Another woman, a street vendor, asks for some anti-odor spray, not because the toilet stinks, she tells me, but because it would be a nice perk. Then we come across sixteen-year-old student Lovencie Pierre, who is hanging out with her friends while her mother is out working in one of the country's ubiquitous lottery stores. Welcoming the novelty of foreign visitors, she leads us through an alley with several inches of water. Kramer wades through in her flip-flops, while the Haitian photographer on my team tiptoes in his pristine sneakers. I'm wearing old sneakers, which I don't care about, but I wish I had rubber boots.

At Pierre's home, the benefits of the SOIL toilet become clear. The toilet is usually in a spot outdoors, she explains, but when rain threatens they pick it up and move it inside. Now it sits in an unfinished room, between a rusty folding cot and a big old refrigerator. Sometimes this room will also flood, she says, in which case they raise the toilet onto a table until the waters drain. Using the toilet wasn't comfortable at first, she admits—she would ask her mother to cover her poop with peanut shells and bagasse. "But now I'm getting used to it," she says. "It's okay."

In a courtyard outside Pierre's house I see the remains of her family's old pit toilet. It filled up, Pierre explained, so they had it emptied, but that involved demolishing the structure around it. They couldn't afford to build a new structure, so they stacked metal and wood on top of the open hole to close it up and instead now pay the monthly fees for the SOIL toilet. In the meantime, across the alley, a neighbor's latrine has no seat or cover, just a broken piece of concrete to sit on. The pit is filled nearly to the rim with a watery mix. With the high water table, Kramer says, "you're just pooping into the groundwater." There's no city water service, so some 60 percent of the population uses well water for washing and other domestic purposes and 37 percent uses it for drinking, so pit latrines like these put them all at risk. The neighbor uses this toilet because it's what her landlord provides. She would be happy to have a SOIL toilet instead, she says.

SOIL serves more than a thousand households and continues to steadily grow—although if people can't pay, the team takes the toilets back. Kramer's goal is to scale up quickly to serve some eight times the current number of customers, but that will require new infusions of funding. SOIL has shown that even the poorest will pay something for toilets, but the costs of collection and treatment are just too high: it would not survive without outside support.

After the collection team picks up the full, covered containers in three-wheelers (or, for narrow streets, in wheelbarrows), they take them to depots to load onto a flatbed truck, which then drives to the composting site some seven miles outside the city. The logistics of pickups and transportation are difficult, Kramer tells me, so they are working with a data-science nonprofit to develop a software tool to create more efficient routes. Fuel is their biggest cost, and when the price has risen, it has at times in turn contributed to another problem for them: social unrest, which can stop collections for a few days altogether when people block the streets.

At the composting site, the labor looks grueling, especially given the intense heat. I watch six strong men empty the pails into bins the size of a horse stall. They wear purple cloth face masks, working in silence to maintain focus. Every once in a while, a newly opened pail stings my nostrils with the whiff of ammonia—the result of users who have let a little pee in. Thanks to microbes that naturally colonize the piles, the mixture in the bins can reach nearly 160 degrees Fahrenheit, enough to kill pathogens. The brownish liquor that runs out the bottom of the bins, which they call *ji kaka*, or poop juice, gets collected and sprayed back on top to encourage the microbes.

After the poop spends about ten weeks in the bins, the workers unload it onto concrete slabs, where they turn the compost every two months for about six months. After that, they shake the compost through a sieve and shovel the fine particles that come out the bottom into bags for sale. The larger particles that get caught in the sieve are fed back into the process, serving as bedding and cover material for the piles in the bins. The more than forty tons of waste that come in per month result in about ten tons of dark brown compost, rich and ready to repair the land. The system is better than pit latrines for the climate, since it has lower emissions of the particularly potent greenhouse gas methane and the compost can sequester carbon in soils.

I talk to the workers at the composting center for a long time. They look hot and exhausted, but they tell me that they find their work meaningful, that they are contributing to making a better country. Traditionally, people who empty pit latrines in Haiti—known as *bayakou*—live with terrible stigma, often doing their work under the cover of night. But these men pose for photographs and walk through the communities in daylight.

A few months later, I see Kramer in the air-conditioned conference center of the Faecal Sludge Management conference in Cape Town. A session on the new night-soil idea overflows with attendees who are eager to learn more about SOIL and similar projects in Ghana, Peru, and Kenya. Together, under the banner of the Container-Based Sanitation Alliance, they have released studies and standards, as well as organized lectures and happy hours. They want experts to consider container-based sanitation as an alternative to sewers and underground pits and tanks, especially in the growing numbers of informal settlements on the outskirts of cities worldwide.

Yet in many places, including South Africa, container-based sanitation concepts still face cultural resistance, and for understandable reasons. In informal settlements, many residents remember the days under apartheid when people had no choice but to make use of twenty-five-liter (6.6-gallon) buckets with toilet seats on top. Today, Cape Town provides another, more hygienic solution to certain areas: regular collection service for twenty thousand plastic flushing camping toilets, nicknamed *porta portas*. To clean them quickly and safely, the city recently built a facility with machines that can do it at five times the speed of the manual cleaning they previously did, handling thousands of the portable toilets per day—making it perhaps the largest container-based sanitation program in the world, though it is not part of the alliance. When I visit (before it fully opens), the whole operation looks something like a high-tech car wash and smells something like a chemical port-a-potty.

On a formal tour of an informal settlement, some residents tell me that the private toilets are a welcome improvement over communal toilets, especially at night. But for many, they are a politically charged reminder that deep racial inequalities in basic services persist some three decades after the end of apartheid and are likely to continue into the future. *Porta portas* and other temporary solutions are "culturally and socially unacceptable, pose health and safety risks, and result in a lack of privacy and dignity," said Axolile Notywala, the general secretary of the Social Justice Coalition, which advocates for permanent infrastructure in settlements, at the Cape Town conference. Because of cultural norms, men, in particular, find using these toilets humiliating, especially since most of the tiny homes don't have a separate space for them, a resident tells me. All in all, they remind some of the old buckets, which are so reviled that the word *bucket*, which sounds descriptive to me, has been nixed by the Container-Based Sanitation Alliance (hence my use of *pail* or *container* to describe SOIL's cylindrical receptacles in this chapter). In the winter of 2013, protesters flung full *porta portas* onto the highway, at the Cape Town International Airport, and onto the steps of the provincial legislature.

In this water-stressed land, however, expanding the conventional flush toilet system would be environmentally irresponsible, even if socially just. Perhaps equity will come from the opposite direction, as people with flush toilets opt out. On my last working day in Cape Town, I attend a planning meeting of a small group that would like to move to off-grid, waterless toilets, after having been shocked by a

2018 drought, when the city almost ran out of water. "If we go through another Day Zero, there's going to be a bigger push. And it's going to be global," one attendee says. To my surprise, many of them believe that well-off people will willingly subscribe to a SOIL-like service that makes products from the toilet waste. And perhaps they're right: the temptation to skip the pipes will only get stronger, as cities struggle more and more with their aging, leaky, and wasteful sewers.

Taking the Piss

We're going to need more pipes.

"There's no use in denying it: this has been a bad week.
I've started drinking my own urine."

—Bret Easton Ellis,
American Psycho (1991)

Urine Heaven

I was eight weeks pregnant, and I hadn't told a soul but my husband
and my doctors. And then I opened up to a kind, middle-aged Dutch
lady named Wilma. Sitting across from me at my dining-room table, she
explained how I would hoard my urine for the next two months. First,
I would pee into a beige plastic pitcher, which I would then pour into
a 1.5-liter (1.6-quart) blue plastic screw-top jug. Every Wednesday, a
driver would come in an unmarked van to collect a crate of six full jugs
and drop off empty ones. From that urine, a pharmaceutical company
would extract a pregnancy hormone, called human chorionic gonadotropin (hCG), that can stimulate ovulation, helping other women get
pregnant, too. It would be discreet, Wilma promised—though, later,
neighbors told me that they noticed.

The newly pregnant author outside her home in the
Netherlands with jugs from Mothers for Mothers.

A pitcher and jug from Mothers for Mothers.

They knew the meaning of those blue jugs because Dutch women have been doing this transfer since 1931. The organization Moeders voor Moeders, or Mothers for Mothers, is just a remnant of what was once a big business of extracting hormones from pee. At first, pharmaceutical companies set up collection programs in Europe and Israel; one even tapped postmenopausal Roman Catholic nuns at an old-age home in Italy. But as worldwide demand for urine-derived drugs increased with the rise of fertility medicine, operations moved to areas with higher populations and birth rates, such as Korea, China, India, and Brazil, though those programs struggled with concerns about consistency, purity, and safety, as well as criticism that they were exploiting the poor and brown on behalf of the wealthy and white. Then, in the 1990s, scientists figured out how to make the hormones in a laboratory, shutting most of the collection programs down, but Mothers for Mothers continues its operations in the Netherlands.

Today, about thirty-five thousand people donate nearly 350,000 gallons of urine to the organization every year. In return for the donation, the donors get nothing but a symbolic gift—I received an impossibly soft baby blanket—and the gratification of being part of what cultural analyst Charlotte Kr[ø]lokke of the University of Southern Denmark calls an "intergenerational Dutch success story, tying the previous donations of older women together with the current donations of younger, pregnant women." And, to keep the chain going, women like Wilma sit at table after table, convincing women like me to keep their pee out of the toilet.

Before flush toilets, many human societies were loath to waste urine. They would collect it, either at home or in public, via urinals or jugs left on the street. One common use was fertilizer, since, more than poop, it contains an array of nutrients that plants need. When stored, urea—the main, nitrogen-rich compound in urine—breaks down into ammonia, the gas that gives urine its characteristic odor. Stale urine has a high pH, which helps to kill off pathogens. Although he couldn't have known the details of this chemistry, ancient Roman writer Columella recommended applying stale human urine to pomegranate trees to make them more fertile and improve the taste of the fruit.

Aged urine's high pH also made it a popular substance for loosening fat, hair, and flesh from animal hides, as well as for softening the hides, in preparation for tanning. Also for cleaning: ancient Roman laundries (called *fullonicas*) would pour urine into vats of dirty clothes, where workers would stomp on them. Indeed, urine was so valuable that the emperors Nero and Vespasian levied a tax on public urinals, called the *vectigal urinae*. According to the Roman historian Suetonius, Vespasian's son Titus found the nature of the tax disgusting, so Vespasian "held a piece of money from the first payment to his son's nose, asking whether its odor was offensive to him. When Titus said 'No,' he replied, 'Yet it comes from urine.'" This is the origin of the term *pecunia non olet*, meaning "money doesn't stink." It is also the origin of the French term *vespasiennes* for certain types of public urinals.

In medieval and early modern England, families collected their urine in clay pots. Artisans used this "chamber lye" for felting and as a "mordant" to set plant dyes, preventing the color from running out when the freshly dyed fabric was next washed. It also helped in the production of alum, another fixative for the textile industry, and alum quarries on the Yorkshire coast imported urine from locations as far-flung as London and Newcastle. Urine, poured on mixtures of manure, ash, and other substances, could form potassium nitrate, or saltpeter, by chemical reactions involving the nitrogen in the pee and the potassium in the ash. Saltpeter is the main ingredient in gunpowder.

People also collected urine for personal-care products and even medical treatments. The ancients made mouthwash and toothpaste from it. Wrote the poet Catullus to the ever-grinning Egnatius, insultingly: "The fact that your teeth are so polished just shows you're the more full of piss." The Swedish, among others, once applied urine to wounds and dry skin, and soldiers irrigated battlefield wounds with it, which is not a great idea since scientists discovered a few years ago that urine is not sterile, as was once believed (although in some situations it might be less contaminated than the available water).

More recently, the singer Madonna has claimed that she prevents athlete's foot by peeing on her feet, which is remotely plausible since factory-made urea is a skin softener and exfoliant used in athlete's foot and dry skin treatments. In one episode of *Friends*, Chandler pees on Monica after she's been stung by a jellyfish at the beach; although scientists disagree with this advice, the laughter that results might relieve some of the pain. And soon there may be new medical uses for urine: in advanced labs, scientists are harvesting stem cells from urine, a step toward more personalized medicine.

Some call urine nature's elixir, and the Indian Sanskrit text known as the *Damar Tantra* claims that raising a yellow glass can cure any disease. But be careful, as there are risks to drinking too much of this beverage, which is why an Army Field Manual lists urine on its "DO NOT drink" list, though people have doubtless tried it under duress, like a trapped Chinese man who survived for six days after a 2008 earthquake, in part by drinking his own urine. Writes a journalist in *Slate*:

> When you drink your own pee, all the stuff that your kidneys had attempted to excrete comes right back into your stomach, and much of it ends up back in your kidneys. After several days of this, your urine will become highly concentrated with dangerous waste products, and drinking it can cause symptoms similar to those brought on by total kidney failure. At that point, you're doomed either way—from dehydration on the one hand or renal meltdown on the other.

Today, people are keen on "pee-cycling" again—possibly more than ever. Not only would it make use of urine, but it would also save the water that would otherwise get used to flush it and prevent its nutrients from becoming pollution. But there's one problem: my urine donations aside, most people probably aren't going to start peeing in jugs and pots again.

Drinking from the Bowl

Every day, according to one estimate, people worldwide use almost 40 billion gallons of freshwater—nearly six times the daily water consumption of the entire continent of Africa—to flush toilets. We can do better. Low-flow toilets got off to a rocky start in the early nineties; due to poor design, they would clog more easily, and often people would flush them several times, defeating the purpose. Since then, however, certain good low-flow technologies have brought down flushes from 3.5 gallons to under today's American standard of 1.6 gallons. And they could go lower, by forcing the poop out with air pressure, applying a super-slippery Teflon-like coating, or making use of rainwater or gray water from showers and washing machines. New ideas might even make it possible to do without water at all. One innovative toilet, from scientists at Cranfield University, scrapes poop out like a rubber spatula in the bowl of a mixer and then incinerates it.

But low-flush innovations can't change the underlying nature of conventional infrastructure, which needs water to work. The Germans practice water saving with fervor, though they live in one of the most water-rich countries in Europe. But the loss of water to the system has led to stagnating, stinking sewers in cities including Cologne and Berlin, which then have to spend money to fight rotten-egg odor and corrosion caused by sewer gases, as well as on awareness campaigns to encourage people to, you know, feel free to take longer showers. Sometimes they even have to pump water through. The same plague struck San Francisco, helped along by a city-funded rebate program for low-flow toilets, leading to hundreds of millions of dollars in upgrades and odor-fighting measures, like bleach.

In South Africa, where sewers are a British colonial legacy, one expert estimates that more than half of the drinking water goes toward sewage management. In 2018, residents of Cape Town managed to cut the city's water consumption by more than half during the Day Zero crisis that threatened to dry up the taps completely. They stopped watering their gardens, limited showers, and started letting "yellow mellow." It was hailed as a conservation success story, but for city wastewater operators it was a nightmare. During the drought, there was an enormous increase in costly blockages because people had not used enough water to wash buildup through. The city even considered calling for a citywide group flush, sometimes called the royal flush, at predetermined times—say, between six and eight in the morning and again later in the day—that would provide enough water flow to move all the fecal material to the treatment plants. And, when the winter rains came, all the built-up sewer gunk came rushing down into the city's wastewater treatment plants. Residents heaved a sigh of relief, but the operators couldn't. It was "extremely stressful," one told me the following year. Cape Town had saved its water—but, unbeknownst to many, nearly broken its sewage infrastructure.

If we can't stop making wastewater, we can at least reuse it. Today, according to one estimate, humans produce about 100 trillion gallons of wastewater annually worldwide, which is about five times the amount of water passing over Niagara Falls annually, and enough to fill Lake Ontario in four years. And, as populations grow and people get connected to piped water, scientists predict the amount will rise by nearly a quarter by 2030 and by half by 2050. By then, about half of the world's population will live in water-stressed regions, where demand for water at least sometimes exceeds supply.

Wastewater reuse already happens unintentionally all the time. In much of the world, farmers often grab whatever water is closest to them, without reference to where it comes from and whether it is treated. People also bathe and wash in fecal-contaminated rivers, as well as take water from them for drinking, cooking, and other activities. Needless to say, this can be dangerous. But reuse has also been done safely for decades. Sometimes it requires new technologies, but more often the hurdles lie elsewhere—in the economics, in the psychology, and in the pipes themselves.

Israel provides one model: the dry country pipes treated city waste-water to agricultural areas, reusing about 85 percent of it for irrigation. Israel's system works in part because the country is small and agri-cultural areas are close to the cities that the wastewater comes from. What's more, Israel prices water so that it makes financial sense for farmers to use treated wastewater instead of groundwater.

Industrial plants can also reuse treated municipal wastewater. In the south of the Netherlands, Dow Chemical Company's Terneuzen manufacturing campus—acres and acres of labyrinthine metal pipes, steaming columns, and utilitarian buildings—turns hydrocarbons into the chemical building blocks for plastics and other products. It needs lots of water for processes that require steam for heating, as well as for cooling towers. You might think that any water would be suitable for this "process water"—after all, nobody drinks it—but, surprisingly, it must be filtered to levels even purer than drinking water or else deposits start to clog up and corrode the equipment, kind of like the scaling in an electric kettle. In order to reduce its dependence on freshwater from a nature area far to the north, the company decided to tap the local wastewater treatment plant for some of its process water, purifying it with an advanced system. It helped that they had found some old, unused pipes that already ran between the city and the facility, which they could repurpose. The project made economic sense and also helped the company to improve its less-than-stellar environmental record.

Cities can also use treated wastewater within their boundaries for "non-potable reuse," such as irrigation of golf courses and highway medians. This water requires a separate set of pipes. In the official system, a blue pipe is for potable water; green is for sewers; yellow is for natural gas, oil, petroleum, or other potentially flammable substances; orange is for telecommunications; and red is for power lines. As for non-potable water, in the 1980s a severely color-blind engineer at the Irvine Ranch Water District in California declared that lavender stood out to him more than any other color. Eventually, the industry adopted this eighties shade of purple as the official color of non-potable reuse systems, helping to ensure that a careless or confused plumber won't accidentally send non-potable water to, say, drinking fountains or showers. Today, cities are laying down more purple pipes and expanding uses to public parks, city buildings, schools, car washes, firefighting, building cooling—and, of course, back to toilets themselves.

Or we can just agree to drink our wastewater. As I was writing this book, I learned that I've sipped recycled wastewater—indeed, I was practically raised on it. In Northern Virginia, where I grew up, drinking water came from the man-made Occoquan Reservoir. When the reservoir was first built, the region was relatively rural, but by the early 1970s, the suburbs had exploded, resulting in eleven wastewater treatment plants discharging to tributaries that flowed into the reservoir. In dry years, 80 percent of the water flowing from the reservoir into the drinking-water systems was treated wastewater. The region was unintentionally operating what's called an indirect potable reuse system. So, in 1978 (just in time for my birth), the authorities opened a state-of-the-art wastewater treatment plant some six miles upstream from the reservoir, in which they treated all the sewage to nearly drinking-water standards before discharging it into the reservoir. Downstream, a water treatment plant treated it again before serving it to houses like mine.

Today, the most innovative, long-term indirect potable reuse program in the United States runs in Orange County, California, just thirty-five miles from Los Angeles. In the early 1970s, the fast-growing county was contending with an overpumped coastal aquifer, which meant insufficient and increasingly salty groundwater, leading it to import freshwater from Northern California and the Colorado River. The water district responded by injecting treated wastewater, mixed with groundwater, into the aquifer at the points where seawater was threatening to flow in, which both protected and replenished the aquifer's freshwater. This water needed to be treated to a high level of purity so as not to clog the recharge wells or add too many salts to the aquifer. So the treatment plant became the first major wastewater-recycling project to use a process called reverse osmosis. Pumps force the dirty water through a thin sheet with tiny pores in it that are just larger than a water molecule, so that the impurities stay on one side and pure water comes out on the other. Despite high costs, the recycled water is still cheaper than water piped in from the north.

Orange County's technology is so good at cleaning the water that it could probably pipe the water directly from the wastewater treatment plant to its drinking-water treatment plants and then straight to people's taps, in a continuous loop. This is called direct potable reuse, and it's still pretty rare. Texas is one state that permits it, and a couple of towns have turned to it during cruel droughts. In 2013, the town of Big Spring, which doesn't get water from a spring but some man-made lakes, became the first American city to adopt direct potable reuse, sending treated wastewater, which is then mixed with raw lake water, to water treatment plants. The wastewater treatment includes microfiltration, reverse osmosis, and ultraviolet disinfection.

Still, one of the biggest hurdles to potable reuse is getting people to accept it—this is known in the industry as the "ick factor." As behavioral scientist Carol Nemeroff put it to the journal *Science*: "How do you get the cognitive sewage out, after the actual sewage is gone?" Regardless of the number of filters involved, people find the idea of drinking reclaimed wastewater pretty gross. In Big Spring, the city's iron pipes mean that the water comes out of the taps brownish anyway and most people use water filters at home, so people accepted the change. But in Australia, a town called Toowoomba started to install a recycling system during a major drought but dropped it, at least temporarily, when the town acquired the unfortunate nickname Poowoomba. And, in 2017, for an outreach program, Orange County bottled its recycled water and passed it out for free, starting with the streets of Hollywood on a hot day—to mixed results. Some took it in stride, but one resident, speaking to a local news program, declared, "What do I look like, a dog or something? I'm not drinking no toilet water!"

One path to acceptance, perhaps, is to turn wastewater into another, more fun kind of beverage. It takes three to seven gallons of water to produce just one gallon of beer. In 2018, a Swedish brewery, in collaboration with beer producer Carlsberg and the Swedish Environmental Research Authority, produced a "clean, crisp" limited-run pilsner called PU:REST (one potential pronunciation, I assume, is "*poorest*"), which they claim sold well at festivals and in stores. And, around the United States, the Pure Water Brewing Alliance has held recycled wastewater competitions for home brewers and craft breweries. Mobile units collect water downstream from wastewater treatment plants and treat it with reverse osmosis and other filters to remove all impurities, turning it into a tasteless "blank slate" that is better for the brewers than tap water. Plus, there might just be a sort of circular logic at work. After all, more beer tends to yield more flushes, so more flushes might as well yield more beer.

Mayor Pete's Smart Sewers

When Pete Buttigieg, who stepped into the national spotlight with his 2019 Democratic presidential primary run, took over as mayor of South Bend, Indiana, in 2012, he inherited a lot of problems—a decaying city center, rampant brain drain, and a problematic police force—but there were also the regularly overflowing sewers.

If the original sin of toilet systems is that they put everything down one hole, *combined-sewer* systems take it to the next level, adding stormwater that runs off streets and buildings to the mix, instead of channeling it into stormwater-only pipes. Combined sewers have pluses: Chadwick, the Victorian sewer designer, favored them because it's less expensive to build one sewer instead of two and the stormwater helps push the wastewater through, cleaning out blockages. South Bend's original combined sewers channeled sewage and stormwater straight into the St. Joseph River—no muss, no fuss.

In the 1950s, the city decided that it didn't want sewage in the river anymore and connected all of the combined sewers to an interceptor pipe that parallels the river and terminates in a treatment plant. That, however, didn't eliminate the pollution. That's because if a storm (or something else, like a blockage) causes a greater flow than the pipes can handle, most combined-sewer systems discharge the excess into waterways, in what's called a combined-sewer overflow. Without that, sewage would stream out of manholes and back up into basements, and treatment plants would stop working due to overloads.

South Bend is not unique. Today in the United States, combined sewers serve about 40 million people in about 860 communities, mostly in the Northeast and Great Lakes regions, according to the EPA, and other countries also have them. Engineers designed combined sewers to handle between only two and four times the average sewage flow during dry weather, as estimated at the time of construction. So overflows, which violate the Clean Water Act, happen shockingly often—in some places, every time there's any kind of downpour. New York City's sewers overflow when rains come down at just a tenth of an inch per hour. In 2014, states reported 22 billion gallons of untreated combined sewage discharged into the Great Lakes, according to the EPA, and some data was missing for many states.

Big rainfalls can close beaches, while major storms can cause massive failures, as fecal matter floods out of sewers, contaminating streets, homes, businesses, wildlife preserves, and waterways. "Florida's Poop Nightmare Has Come True," read a headline after Hurricane Irma in 2017. There's sometimes a basic injustice at play: in some American cities, overflow points are disproportionately found in low-income, nonwhite communities, forcing the vulnerable to endure not only their own sewage but also that of the privileged and well-heeled.

As unpleasant as combined-sewer overflows are, the menu of options for fixing the problem is unappealing in a different way. One possibility is to dig up the combined sewers and separate them into sanitary sewers for wastewater and storm sewers for stormwater. This is sometimes the cleanest option, but it's disruptive and expensive and doesn't take into account the pollution in stormwater itself. Plus, sanitary sewers can and do also overflow due to leaks and other failures. Instead, some big cities have built extra storage, where sewage can ride out the storm before getting directed back to the treatment plant. Washington, D.C., recently built an enormous tunnel system to hold overflow that would otherwise pollute the Anacostia and Potomac rivers. To pay for it, the utility issued a $350 million green century bond, an innovative financing tool that will spread the costs out over the next hundred years, passing them along to all of the people, some still unborn, who will benefit from the works. And Chicago's Deep Tunnel, already partly built, aims to have 17.5 billion gallons of storage space by 2029. Yet even this isn't enough for extremely heavy rainstorms—the kinds that, in an era of climate change, seem to surprise us more and more.

Down-and-out South Bend, a city of about one hundred thousand, could hardly afford projects like these. But rather than pay massive fines to the EPA for overflows, Buttigieg's predecessor signed a "consent decree" that committed the city to about 700 million dollars' worth of upgrades over two decades, writes Buttigieg in his memoir, *Shortest Way Home*:

> That meant almost ten thousand dollars for every man, woman, and child—in a city whose per capita income in 2017 stood at $19,818. . . . Worse, the models showed that the highly expensive plan wouldn't actually achieve the level of control and environmental improvement that was intended—we could do the whole thing and still be in violation.

But Buttigieg had reason to hope that this giant outlay would not be necessary. For years, engineers had been developing alternatives to big digs such as these; it was just that the EPA, as well as many in the water industry, hadn't embraced them yet. One option is a category of solutions called green infrastructure, which captures stormwater in constructed wetlands, rain barrels, green roofs, patches of vegetation, and pavement that absorbs rain. These prevent the stormwater from entering the sewers in the first place and serve the additional purpose of directing the water to nature. Another option might be to put small, possibly mobile, treatment plants at overflow points. A Milwaukee-based company called Rapid Radicals, headed by engineer Paige Peters, is developing a technology that uses chemical reactions to treat flows in thirty-five minutes—far quicker than conventional microbe-based wastewater treatment, which takes about a day.

In South Bend, Buttigieg's predecessor had laid the groundwork for another solution that's made it a pioneer in the sewer world. Starting in 2004, the city partnered with civil and electrical engineers at the University of Notre Dame, which is located in South Bend, and later with Purdue University, to develop a new technology: wireless sensors that could monitor and respond to water levels throughout the six-hundred-mile wastewater system. Placing sensors in the harsh sewer environment is a challenge, as they have to survive high humidity, fast-moving debris, and slime. South Bend's are housed in explosion-proof boxes because of the risks from sewer gases. Today, the network of more than 150 sensors and other connected devices, sharing data every five minutes, helps detect blockages and potential overflows—something previously done by workers who would drive around the city for forty hours per week, lifting manhole covers to check for rising water levels or other signs of trouble. Connected control valves, gates, and movable weirs open and shut to redirect water.

Importantly, the system can decide for itself in real time how to best use the infrastructure that's there. Early in the project, researchers realized that the system wasn't always using the full capacity of the interceptor pipe, which transports sewage and stormwater from the combined sewers to the plant. If it rained heavily in just one part of the city, the interceptor couldn't adjust to allow more water in from that section, and so the sewer would overflow into that part of the river. But with the new network and some added hardware, the system can now ensure that the interceptor always runs as full as possible. Still, overflows can happen when there's just too much rainfall.

Notre Dame spun out a private company called EmNet to commercialize the technology and take it to more combined-sewer cities like Elkhart, Indiana; Hoboken, New Jersey; and San Francisco; it was so successful that the global water technology company Xylem bought it in 2018. For all the benefits of connected sewers, however, there's a potential downside to making any critical infrastructure "smarter": cyberattacks. Some American water utilities have been held ransom, taken down for days, or even had their water flow manipulated. Recently, Israel reported a major attempted, but unsuccessful, cyberattack on the control systems of wastewater treatment plants, pumping stations, and sewers.

For South Bend, the result, after about a $6 million investment, as well as some other improvements, was that it reduced its combined-sewer overflows—from 1 to 2 billion gallons a year, depending on the rainfall, to less than 500 million gallons most years—giving it a strong argument for renegotiating the terms of its consent decree with the EPA. It presented a revised proposal, costing about a third of the original, involving fewer underground storage tanks and more green infrastructure such as rain gardens, rain barrels, and porous pavement to catch stormwater. The data from the sensors helped them model how those changes would play out.

And the international recognition that this project has received has made Buttigieg, who stepped down as mayor in 2020, one of the few politicians who will enthuse about sewage on the campaign trail. "What makes a country great isn't chauvinism. It's the kinds of lives you enable people to lead," he told *Rolling Stone* magazine, in a seeming rebuttal to the Make America Great Again ethos, during the presidential primary campaign. "I think about wastewater management as freedom. If a resident of our city doesn't have to give it a second thought, she's freer." South Bend, he brags, has the "smartest sewers in the world"—if, I would add, you can ever call a sewer that mixes everything together smart.

Yes Wee Can

The restrooms at Wageningen University's Environmental Technology Department in the Netherlands double as engineering laboratories—you can't be quite sure what manner of toilet you'll find behind each door, whether it will work, and how it will smell. I try a urine-diversion toilet, which has an extra little depression in the front of the bowl to collect pee. It looks kind of weird, something like the examination shelf in German toilets but on the other end. And it doesn't seem all that effective. After all, for women, pee isn't really easy to aim, so some of mine goes into the wrong bowl. For men, it can get even weirder: in some designs, the raised portion that divides the pee and poop sections sometimes accidentally touches men's genitalia. Some researchers have taken to calling this engineering blunder the ball ridge.

Of course, it's not as extreme as the situation in space, where astronauts must separate their streams in zero gravity. On the Space Shuttle's Waste Collection System, astronauts would pee into a personalized funnel attached to a suction hose and poop through a four-inch hole, like a beanbag toss at a carnival. To make sure they could do it, they trained on a mock-up at the Johnson Space Center in Houston, which had a closed-circuit camera inside the seat for practicing alignment. "I think of Peter Fonda in *Easy Rider* riding a chopper," NASA astronaut Michael Massimino once said, revving an imaginary motorcycle. "That's the right position for me."

Some of the most exciting innovations in sanitation depend on being able to separate pee from poop and water at the outset. But these projects all run up against the same problem: dividing the streams is not that simple, and, despite decades of efforts, coming up with a design that doesn't make the user squirm has been essentially impossible.

Take the infamous case of an innovative housing project in the city of Erdos in the Inner Mongolia Autonomous Region of China. It was a cooperation between the Stockholm Environment Institute, the district government, and a real-estate developer. Installing waterless, urine-diversion toilets in the building seemed to make so much sense: when the project began in 2003, a drought meant that many residents got water only three times per day for about an hour, so flush toilets were a waste; the agricultural region had use for urine-based fertilizer, and the Chinese government was committing to a recycling economy.

Yet by January 2009, when the drought had let up, the residents of this three-thousand-person eco-town were lobbying for "normal" flush toilets, and by December of that same year they had them.

What went wrong? Construction mistakes, poor maintenance, odor, lack of local expertise, and failing political support. But the "weakest link" in the project was "household acceptance," a postmortem report reads. Like the urine-diversion toilet that I used, the bowl had two receptacles: a large one in the back for poop and a small one in the front for pee. Since there was no water in these toilets, people had to dump sawdust on top of their poop and then open a chute to drop it down into a pipe. Not surprisingly, "a common comment from households was that the toilets were awkward to use, and explaining their function to visiting family relatives and friends was considered an embarrassment and unnecessary burden." Instead of taking pride in their cutting-edge eco-friendly community, the residents were ashamed of their weird toilets.

But we needn't despair. Change in bathroom fixtures is possible, if often slow. In the nineteenth century, the rise of public urinals gave men an alternative to exposing themselves in public and leaving a stink in every alley. Recently, urinals have been undergoing transformations—of which, as a woman, I was once unaware. It's now common for men to target their streams at an image of a fly placed near the base of the bowl, since otherwise they apparently have a hard time remaining focused. And, more recently (and perhaps in some opposition to the purpose of the fly), TV screens embedded in the top of the urinals play advertisements. But arguably the most important advance in the past three decades has been the waterless urinal.

Normal, water-based urinals work much like regular toilets; the water in a U-bend forms a seal preventing sewer gases from escaping. A flush washes the urine through the U-bend and renews the water seal there. Old models used around five gallons per flush, as much as a normal toilet. In waterless urinals, on the other hand, a special liquid sealant floats over the drain. Lighter than urine, it lets pee down the pipe but no sewer gases up. In essence, gravity, not the force of a flush, does the work.

German entrepreneur Klaus Reichardt first brought waterless urinals to the U.S. market in the early 1990s, making substantial improvements over previous iterations in Europe. But they didn't attract mass interest until a decade later, when a company called Falcon Waterfree Technologies came up with a novel financial model: since the liquid sealant comes in multiuse cartridges, it could operate like the razor industry, charging little for the urinals themselves but selling the cartridges in perpetuity. From the customer's perspective, water savings could offset the cartridge costs. Falcon won the backing of high-profile investors and advisors, including former vice president and environmental activist Al Gore.

At this point, however, the invention still faced major resistance— not from users, as in the case of urine-diverting toilets, but from plumbers. Falcon needed the plumbers' support because the nation's model plumbing codes hadn't taken the possibility of waterless urinals into account and therefore essentially forbid them. But the plumbers opposed amendments to the codes, arguing that sewer gases could be released during cartridge changes, potentially causing an array of alarming effects, including "unconsciousness, respiratory paralysis, and death" (none of which ever materialized). Falcon, for their part, claimed the plumbers had another reason for opposition: since waterless urinals didn't require water pipes, they would result in less plumbing work.

Ultimately, the two sides compromised. Changes to the codes in the late 2000s permitted waterless urinals, but only if there were also water pipes in the walls, in case building owners ever wanted to return to flush urinals. Pressure to save water has only grown since then, though, so waterless is here to stay.

*   *   *

Of course, urinals fail to capture the pee of about half the population—and why should we be left out? When I was young and going out on the town, women always joked that, no matter how much pressure you felt in your bladder, you shouldn't "break the seal"—meaning, that once you peed for the first time, you'd have to pee regularly, leaving you standing in lines for the ladies' toilet while watching men flit in and out of their restroom. Today, at outdoor festivals, many young women report defying gender expectations to squat in the same places that men freely pee—along fences—although they also report getting heckled and punished far more than men for doing so. To them, I say: Smash the patriarchy! Let it flow! But I also say: give these women some urinals.

Copenhagen-based architects Gina Périer and Alexander Egebjerg, founders of Lapee, a female urinal unit that debuted in 2019, agree. "It's so degrading for women, the situation of urination," Périer has said. The team is just the latest to attempt to design a female urinal. One previous model was a seatless toilet that a woman could stand or squat over; another was a tube attached to a funnel that a woman would hold to her crotch. But, as an anecdotal measure of success, I, for one, have never seen one of these installed in a public restroom. To be honest, looking at photos, I suspect it would take practice—and possibly specially designed clothes—to use one without wetting oneself or revealing far more than someone with a penis performing the same activity.

And, for many women, peeing is not easily separable from pooping or menstruating; it's not as straightforward as flexing different muscles. In the words of Carol Olmert in her book *Bathrooms Make Me Nervous*, "To urinate from a standing position, women must undergo a complex sort of retraining, one that reverses or subverts the socialization process." The act also has a culture war aspect to it, writes law professor Mary Anne Case:

> If devices designed to encourage a woman to urinate more like a
> man ever were to catch on, they might themselves generate cultural

anxiety, as the toilet scene in the film *The Full Monty* indicates. In it, unemployed steelworkers spy on their wives, who have taken over the local Workingman's Club for an evening of entertainment by male strippers. The steelworkers come upon women occupying the men's room, cheering on one of their number as she hikes up her skirt and directs a stream of her urine into a urinal. Already threatened in their masculinity, the men conclude, "when women start pissing like us, that's it, we're finished, extincto. . . . They're turning into us. A few years and men won't exist, except in zoos. . . . I mean we're not needed no more, obsolete, dinosaurs, yesterday's news."

Still, Périer and Egebjerg are trying their hand at it. Lapee (as in *la* + *pee*; Périer is French), which can be delivered to outdoor events such as festivals, has three private stalls arranged in a clover shape that ensures that the entrances are not readily visible, although there are no doors. Inside each stall there's an oval receptacle with a drain at the bottom, which women can squat over, their bottom halves blocked from view while their heads peek out over the top like meerkats, allowing them to monitor their surroundings or chat with other users. It's cleaned like a standard four-man urinal, which the founders say makes it easier for contractors to deploy.

It's a strange enough experience that "you need to be a little bit drunk to do it," an attendee at Denmark's Roskilde Festival told a reporter after using it. But at least it's better than squatting by a fence. And, since the company hopes to use the urine from its nearly three-hundred-gallon tanks for fertilizer and electricity, the Lapee will also soon allow women the equal opportunity to donate their pee for use as fertilizer or energy. Perhaps for some nervous women, that could be the incentive to break the seal.

The real breakthrough technology, though, would be a urine-diverting toilet that anybody would want to use, since that would enable building-scale projects, including residential housing, such as the one that failed in Erdos. Harald Gründl, the co-founder of the über-cool, award-winning EOOS design firm in Vienna, Austria, started working on toilets in 2010. For fifteen years before that, the firm had focused on high-end product, furniture, and retail-space design for clients like Adidas and Armani. But Gründl—an understated fifty-something who, if EOOS were a band, looks like he would be the bassist—had the urge to work on something more meaningful that would contribute to the future. After he reached out to the sanitation teams at the Swiss institute Eawag, they decided to collaborate on a design for the first-ever Gates Foundation Reinvent the Toilet Challenge.

The team set out to design a toilet that separated urine and feces so that they could be processed more efficiently. Gründl also wanted it to be a prestige object, something to show off to neighbors, styled with "a bathroom 'look,'" even if it was installed in a shed. And, although it would be a dry toilet, in the sense that it would not add water to the pee and poop, it would let people use water to wash and flush, as a matter of dignity.

The result, after several design iterations, was the Blue Diversion, which was tall, narrow, and the blue of a clean swimming pool on a cloudless day. It was a squat toilet, since that's what the people in focus groups in Kampala, Uganda, preferred. Users either pooped into a hole at the back or peed into a sort of half funnel at the front. When they were done, they could squeeze a small spray nozzle to clean their bums, flush the pan clean with 1.5 liters (1.6 quarts) of water, and wash their hands in a small sink. The water from the washing and flushing got filtered, cleaned, and recycled back through the system, while the urine and feces could be processed right there, if connected to an off-grid technology, or transported to a treatment center.

But there was a problem. When the team took the Blue Diversion to Nairobi, Kenya, for field testing during a very wet season, users tracked in mud on their shoes. This mud messed with the valve system that separated the urine from the water stream, such that all of the wash water flowed into the urine container, leaving no water in the system, in what Gründl calls an "unfortunate chain reaction." The Blue Diversion, they had to admit, was not ready for prime time.

Gründl went back to the Gates Foundation for another grant on just stream separation. At first, the team planned to add a sensor that could tell the difference between urine and water, which would solve the problem that they had faced in Kenya. At the last minute, however, the sensor developer pulled out, spooked by the Gates Foundation's "Global Access" requirement, for which grantees must promise to develop versions of their technology for low-income contexts at a low price. "So we submitted the proposal without a solution," Gründl says. "And then we got the grant and we were in big trouble."

His team decided to go back to basics, quantifying some of the simplest questions of urination and defecation, such as how much comes out, at what speeds, and at what angles. They installed a thermo-imager in a private area of his firm's office and captured rainbow-hued videos of the dozen members of his design team peeing and pooping. Seated men peed, uncomplicatedly, toward the front of the toilet, making separation relatively simple. But seated women's pees were more complex. In one thermo-imaged video, a little cool-blue shower comes down at first, spraying in all directions; then, as the stream builds, it angles slightly forward, turning green and then red and then white from the warmth; finally, it tapers off, dripping straight down. Sometimes, the team found, women's streams overlapped with the feces "fall line."

They incorporated all of their data into a contraption, called the Urinator, which they released as an open-source 3D-printable design. Mounted on a toilet seat, it could "pee" like a man or a woman. This allowed them to experiment with different designs in their workshop, making mock-ups and playing around with the streams to see how they would work.

Eventually—Gründl can't quite remember the specific moment anymore—the team hit upon an idea: you know how tea sometimes clings to the outside of a teapot, dribbling down onto the table, instead of

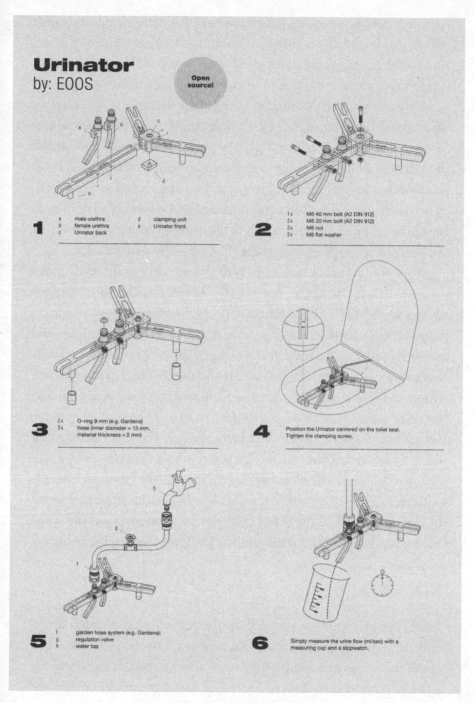

Diagram of the Urinator by EOOS.

pouring cleanly out of the spout into a cup? This is called the teapot effect, and it arises from the phenomenon of "surface tension" in liquids. With the right angles, they realized, they could harness the effect to get urine streams to cling to the inside wall of a toilet.

Gründl was sure that this was the eureka moment for which they had hoped. The team designed what they now call the Urine Trap. In a toilet with this new technology, pee pools on an angled shelf at the front before gliding down the side of the toilet into a hidden hole. Poop plops down into a little bit of water at the back (or into a dry toilet). They even worked with an expert in computational fluid dynamics to make sure they captured as much of the urine as possible—upward of 80 percent, which met their goals. The Urinator confirmed it.

Still, the question remains: Will people accept it? And, more important, is it a design for everybody—not just for the poor? Gründl thinks so. EOOS has partnered with the Swiss bathroom products manufacturer Laufen to incorporate the Urine Trap into a high-end ceramic model, called save!, that's planned for the European market. It has won awards from design festivals and magazines. For low-income contexts, Gründl is also working on incorporating the Urine Trap into lower-cost squatting and sitting toilets as part of his commitment to the Gates Foundation, as well as to a better future. The most astonishing thing about save!, in the end, is what you don't see: an extra bowl, a separate hole, a valve. The metamorphosis is essentially invisible. There is nothing to get used to, nothing to be proud of, nothing to be ashamed of, nothing to talk about—and that may be what most people want. "They just want a nice bathroom," Gründl says. No aiming required.

Pee Is for Precious

I didn't appreciate "liquid gold" until I donated to the Dutch urine-collection organization Mothers for Mothers. Owned by the pharmaceutical company Organon, they were after the pregnancy hormone hCG. It is produced by the placenta, the organ that provides oxygen and nutrition to a growing fetus in the womb. The hormone supports the corpus luteum, a temporary bulge in the ovaries that produces estrogen and progesterone during early pregnancy. After an embryo implants in the uterus wall, hCG doubles every forty-eight to seventy-two hours, peaking at about eight to eleven weeks after conception, which is why Mothers for Mothers collects urine during the first trimester.

Scientists saw hCG's potential as a fertility drug in humans as early as the 1930s, when Mothers for Mothers got started (though it didn't operate under the current name until 1966). In 1950, scientists extracted another useful fertility hormone, human menopausal gonadotrophin (hMG), from postmenopausal women. In 1962, doctors reported the first pregnancy created via ovulation stimulation with hMG followed by ovulation induction with hCG, which later became part of a standard protocol for infertility. Altogether, these treatments have "enabled the birth of millions of children to people affected by infertility," according to endocrinologist Bruno Lunenfeld, a fertility medicine pioneer.

Today, Aspen Oss, a successor of Organon, owns Mothers for Mothers. In her 2018 book, *Global Fluids: The Cultural Politics of Reproductive Waste and Value*, scholar Charlotte Kroløkke describes the operation, after getting the opportunity to visit it. After the driver who picked up my donated urine completed his rounds in The Hague, he would have driven the blue jugs ninety minutes east to a factory in the town of Boxtel for the first purifying step. From there the urine would have continued on for another thirty minutes or so to the small industrial town of Oss, where technicians would have processed it into a powder. (Since then, Aspen Oss has upgraded their Boxtel facility to handle both steps.) The active ingredient derived from the Mothers for Mothers urine would have then been sold on, potentially to the drug company Merck Sharp & Dohme, which makes from it a drug called PREGNYL, used primarily in fertility treatments. Patients who want to get pregnant with their own eggs, as well as those who are donating eggs, inject PREGNYL to trigger ovulation.

*   *   *

Some years ago, Mothers for Mothers tried delivering the liquid left-overs from this process to a pilot plant that turned them into stru-vite, a phosphorus-rich fertilizer. That plant has since closed. It's a shame—while today hCG is one of the most valuable substances that can come out of urine, most everybody thinks that phosphorus will, sooner or later, catch up.

Phosphorus is an essential nutrient in agriculture. It is the P in ATP, the energy-carrying molecule that's formed during photosynthesis and powers reactions in the cells of every living thing. The element also plays a role in building cellular structures and transferring genetic information. Compared to other fertilizer nutrients, commercially available phosphorus is especially troubled. It comes from phosphate rock, which contains heavy metals such as cadmium that can be harmful to the kidneys and bones. And it's mined in often unstable or geopolitically risky places, the most prominent of which is Morocco. Phosphorus producers are secretive, so nobody quite knows how much mineral phosphorus exists around the globe. According to one paper, "its supply chain is a black box," with up to 90 percent of the supply lost "from mine to fork," much of it seemingly ending up as pollution in water bodies, where it can harm ecosystems and even poison people. (Meanwhile, more than half of the global pollution load of phosphorus comes from human waste.)

That leaves many wondering when we will hit "Peak P," which, like *Peak Oil*, refers to the point after which the resource will start becoming harder to get, not to mention more expensive. Controversially, more than a decade ago, some scientists warned that we might see Peak P as soon as 2030. Most put the date later, but, in any case, once it's gone, it's gone, and urine looks to be the next-best source of it, so it makes sense to start developing the tools to harvest it now.

Some farmers in landlocked Malawi, where phosphorus is already at a premium, are among those at the forefront of pee-cycling for agriculture, applying aged urine to crops in an echo of what people did in the past. In Brattleboro, Vermont, the Rich Earth Institute collects more than ten thousand gallons of urine per year from nearly two hundred local residents, pasteurizes it, and then distributes it to four farmers, who drip the liquid onto hay fields. In Switzerland, the company Vuna, a spin-off from Eawag, has taken a higher-tech route, developing a concentrated fertilizer that they make from pee collected in the institute's urine-diverting toilets and at outdoor festivals and markets. Bacteria convert the ammonium in urine to nitrate. This conversion prevents the nitrogen from escaping in the form of ammonia gas. An activated carbon filter then removes pharmaceuticals and hormones, and distillation concentrates the solution into a brown syrup. The end product, called Aurin (a play on *urine* and also on gold's chemical symbol, Au, from the Latin *aurum* for "glowing dawn"), includes all of the macro- and micro-nutrients in urine, including phosphorus, and has been approved by the Swiss Federal Office for Agriculture for use, including, importantly, on edible plants. Bastian Etter, the head of Vuna, gave me a sample bottle when I met him. To use it, I mixed ten milliliters (about two teaspoons) of the syrup with about a hundred times that amount of water and applied it monthly to plants in my house and garden. Unfortunately, it was no match for a late frost followed by an unprecedented drought. But its biggest foe was probably my rotten thumb: though a favorite vine died that year, I cannot blame some of Switzerland's brightest minds, nor their bladders.

Researchers are working on other types of products from urine, too. Environmental and chemical engineer William Tarpeh of Stanford University, nicknamed the Waste Wizard, has harnessed electrochemical processes to extract the nitrogen in urine and transform it into not only a fertilizer but also a disinfectant, which could command a higher price than fertilizer in some markets. From engineer Ioannis Ieropoulos, of the Bristol BioEnergy Centre at the University of West England, there's an idea for how to power up mobile phones using pee-generated electricity. (After all, who has not found themselves simultaneously in dire need of both a toilet and some phone charging?) And from engineer Dyllon Randall, of the University of Cape Town, there are bio-bricks. Typically, bricks are fired in kilns at extremely high temperatures. But Randall adds urine to a mixture of loose sand

and bacteria. These bacteria produce an enzyme called urease, which breaks down the urea in the urine, creating calcium carbonate, a salt. The calcium carbonate essentially cements the sand into place, in a process similar to the way seashells and coral reefs grow. Formed in a mold for various lengths of time, this process can create building blocks of any shape and many different strengths.

According to the National Institutes of Health, people have used urine to check for pregnancy since at least as far back as 1350 BCE. An ancient Egyptian document describes how women could urinate on wheat and barley seeds. "If the barley grows, it means a male child," the papyrus reads. "If the wheat grows, it means a female child." No growth meant no pregnancy. In 1963, scientists tested this and found that it worked pretty well: 70 percent of the time, pregnant women's urine promoted growth, unlike other people's urine, perhaps because of elevated estrogen levels. In the Middle Ages, people called piss prophets relied on signs such as urine's appearance—"clear pale lemon color leaning toward off-white, having a cloud on its surface"—as well as mixing the urine with wine, to reveal pregnancies.

The 1930s saw the introduction of a more scientific approach, known as the "rabbit test." A woman's urine got injected into a rabbit (or sometimes another animal) to see if any hCG in it would induce ovulation in that animal. This gave rise to the gloomy euphemism *the rabbit died*, meaning that a woman was pregnant, though the rabbit always died, since technicians had to kill the animal to check its ovaries. Then a zoologist with the unlikely name of Lancelot Hogben developed the first widely available pregnancy test when he discovered that the female African clawed frog, *Xenopus laevis*, would lay eggs when injected with pregnancy urine, obviating the need for dissection. That ended in the 1970s, but there's an interesting, if sad, side note: the international trade in these frogs was so extensive that it may have managed to contribute to the spread of a devastating amphibian pandemic that has done more damage to global biodiversity than any other disease in recorded history, driving declines in more than two hundred amphibian species. Known as chytrid or *Bd* (for *Batrachochytrium dendrobatidis*), it's a fungal infection that attacks the animals' permeable skin. As one biologist put it: "If it were a human pathogen, it'd be in a zombie film."

Finally, in the 1960s, scientists developed animal-free ways to test for the presence of hCG. Organon produced the first of these tests, called Pregnosticon, for doctors' offices. Later, one of the company's New Jersey–based designers, Margaret Crane, adapted it for home use. The kit had users mix their urine with antibodies that bind to hCG, red blood cells (from sheep) coated with hCG, and distilled water. If the patient was pregnant, the antibodies would bind to the hCG in the urine and the hCG-decorated blood cells would settle to the bottom, forming a distinctive red ring within a couple of hours. If the subject wasn't pregnant, the antibodies would cause the hCG-coated blood cells to clump together and there would be no ring. The hCG for those early kits came from the urine of pregnant women, collected by programs such as Mothers for Mothers. But my urine didn't go to make a home pregnancy test: today, donated urine goes to medicines alone.

Mothers for Mothers carefully cultivates its image, conveyed in the slogan *geluk kun je delen*—you can share happiness. But the reality is not just smiling mothers and babies. In the mid-1980s, Kroløkke writes, donors nearly started a boycott when journalists reported that the program was supplying leftover product for the reproductive management of pigs. Labeled with the nickname Mothers for Sows, the organization ended that practice in the Netherlands, although they continued with it for a while in Brazil. Today, some people follow the so-called hCG diet, which is rife with frightful get-thin-quick promises. Male athletes also sometimes inject the hormone as a testosterone booster, a kind of doping. Mothers for Mothers has publicly expressed disapproval of some of these uses—especially in dieting— Kroløkke writes, but there's not that much they can do about it.

And, ultimately, my generous donation helped Aspen Oss's bottom line. "I didn't do it, to be honest," environmental biotechnology researcher Miriam van Eekert of Wageningen University told me, referring to when she was pregnant. "Part of it was like, okay, now I'm donating my urine to a company and they actually make money?" Though her own research aims to recycle human waste, she doesn't buy what Krol\u00f8kke calls the "collective ethos of patriotic solidarity and sisterhood" that Mothers for Mothers tries to inculcate. Indeed, human geographer Kate Boyer of Cardiff University in the United Kingdom has noted that "corporate third parties are increasingly profiting from biosubstances harvested for free." Mothers for Mothers softens that capitalist reality by making the donation sound like a philanthropic gift.

Curious about women's experiences on the other end of the process, I ask on Facebook whether anyone has used PREGNYL. A friend's friend raises her hand. We meet up, each with a young child in tow, and I smile to see her seven-month bump. She has not had an easy time getting pregnant with this second one, she tells me, and her doctor prescribed PREGNYL as the trigger shot for several fertility treatments, including IVF, although, in the end, she got pregnant "spontaneously" between treatment cycles. She describes the physical pain and the intense emotions of that time; it was hard, she says, to separate the responses to the PREGNYL itself from those to the other drugs and procedures that were part of the treatment. "All of it is pretty heavy on your body," she says.

When I reveal to her that the drug is made from the urine of pregnant Dutch women—including *my own*—she's shocked. "Oh my god, that's crazy," she laughs. But within minutes, reassured of the safety, she's talking sisterhood, too. "It's such a difficult time if you're not getting pregnant—it's emotional; it's stressful; you get pregnant; you lose a pregnancy. So I think that a lot of us who have been through it have a lot of sympathy for other women and would want to do something."

Though I did not use fertility drugs myself, I also struggled to get and stay pregnant. So, looking back, I can't regret my donation. I never found the process dirty or onerous, though some do, in part because the preservatives in the blue bottles emit a distinct smell that can make already nauseated people really want to barf. And though I would prefer that profits benefit our sanitation systems instead of global pharmaceutical companies, it's hard to feel too cynical about "solidarity and sisterhood" with other mothers. I also suspect that this minor act of altruism helped me cope with my own fears around my pregnancy as well as the transition into my new identity as a mom, in which sharing fluids with another human would become routine. As a bonus, I saved all the water that I otherwise would have flushed for those number ones.

After a short while, peeing into a pitcher became second nature to me. So much so that, when it came time to stop, I couldn't help but feel that I was flushing something rare and precious down the pipes.

Eating Sh!t

That sludge isn't going to digest itself.

C'est ici que tombent en ruine
Tous les chefs d'oeuvre de la cuisine.
[Here fall in ruin
All the masterpieces of cuisine.]

—French restroom graffiti,
first recorded in 1899

Black Gold

When Carl Pawlowski was a boy in the 1970s, his hometown of Quincy, Massachusetts, a suburb south of Boston that's the birthplace of President John Adams and his son President John Quincy Adams, was known as the Flounder Capital of the World. An avid boater since those youthful days, Pawlowski remembers the masses of flapping flatfish well: "People would get in buses from New Jersey—buses!—take them up here, rent boats, and fill trash barrels with flounder," he says in his captivating Boston baritone. Locals recall that kids would offer to fillet the catch for ten cents a fish. A nearby cafe would batter and bake them to eat on the spot.

It was fun, sure, but it wasn't a sign of a healthy fishery. The under-funded wastewater treatment plant on nearby Nut Island—together with one farther north called Deer Island—had failed to keep up with population and industrial growth in Boston. "It was a parody of sew-age treatment," says ecologist Anne Giblin, of the Marine Biological Laboratory in Woods Hole, Massachusetts, who studied the effects of wastewater disposal in the harbor and bay for two decades starting in the early 1990s. "They separated the solids from the liquids, and then they put the solids in the harbor in one place, and the liquids in the harbor in another." By the early 1980s, in addition to dumping sludge for the tides to carry away, these plants were also discharging 350 million gallons of untreated sewage every day. Combined-sewer overflows contributed another 3 billion gallons a year. "Boston Harbor was once called the 'dirtiest harbor in the world,'" Pawlowski says, "and rightfully so."

Even young Pawlowski could tell something was wrong. "The water just wasn't clear . . . it was a brownish-greenish type," he says. As a result of all the fats and oils in the discharge, "you could literally see a scum film on the water." The sewage killed off vast meadows of seagrass, which had been important nurseries for young fish and Boston's famed shell-fish. It led to frequent beach and shellfish bed closures. And it fueled an out-of-control food chain, feeding worms that drew mud-dwelling, predatory winter flounder to a few active spots, which delighted the fishermen, except for the uncomfortable fact that the flounder developed liver tumors from exposure to pollutants in the sewage.

But it was the shit that got to people. Famously, Quincy's city solicitor William Golden convinced the city to sue the commonwealth after he took a jog along a beach, slogging through what he thought was seaweed and jellyfish before he realized that it was human feces. The lawsuit was part of a national movement that eventually made it illegal for wastewater utilities to use waterways as a trash receptacle for sewage. Ultimately, a court ordered what came to be known as the Boston Harbor Cleanup.

The new rules, while a huge win for rivers and bays countrywide, created a new problem: Where would the sludge go if not in the water?

Today, the United States produces more than 7 million dry tons of sewage sludge per year. Municipalities have basically three options for that sludge. Some 30 percent goes to landfill. Another 15 percent gets incinerated. But both of those are polluting in their own ways and unpopular with the public, and they're not cheap, since landfill or incinerator space must be secured, and trucking around sludge consumes a lot of fuel.

That's why 55 percent of sludge, including greater Boston's, takes a third route: it serves as a kind of fertilizer called biosolids, powering crops on farms, trees in forests, and grass on lawns. Pawlowski, now an environmental engineer for the Massachusetts Water Resources Authority, plays a crucial role in that route, overseeing the nearly three-decades-old factory, not far from his childhood home and boating spots, that turns Boston's sludge into a sometimes controversial product that he and others in the industry call black gold.

Nobody quite knows when our species started farming with our own poop—any evidence has long since degraded—but the practice probably always served a dual purpose: improving the productivity of fields and finding some place to put all the poop, although it likely also had the side effect of spreading disease. Some cultures venerated shit as manure, while others reviled it, attitudes that are sometimes referred to as fecophilic and fecophobic. In eighteenth-century Japan, where it became a profitable commodity, night soil from wealthier households commanded higher prices, since richer people ate better diets.

In early nineteenth-century Paris, workers dried and aged cesspool waste near the city's dump, transforming it into a gray-black, fine-grained product known as *poudrette*. In the 1830s, as part of his utopian "Circulus" theory, French philosopher and economist Pierre Leroux called for the state to collect human excrement as a form of tax. "If men were believers, learned, religious," he wrote, then "each would religiously collect his dung to give it to the state, that is, to the tax-collector, by way of an impost or personal contribution. Agricultural production would instantly be doubled, and poverty would vanish from the globe."

Soon after, however, European demand for poop-based fertilizer crashed with the introduction of guano, solid bird droppings mined from South America, which was followed by the Germans' invention of factory-made fertilizers, which pull nitrogen gas out of the air. These alternatives had the nutrients but not the inconvenient bulk of manure and increased yields dramatically. But some believed that the use of artificial fertilizers represented a dangerous break in the natural order. The only way to restore that order, these thinkers argued, was to reconnect the broken cycle that linked urban dwellers to country farms through poop. As German agricultural chemist Justus von Liebig wrote to the Lord Mayor of London: "If a possibility is offered to the farmer to get back, as sewage, those matters which he has carried to the town in the form of corn, meat, and vegetables, and if he gives his field the same, both in quantity and quality, as he took from it, then its fertility may be assured for an endless number of years."

Extrapolating from the high price of guano, some proponents proposed wild estimates of the total value of sewage. As sewer systems expanded through Europe, some areas tried irrigating with raw sewage. But pumping the foul mixture to the right places, at the right times, and in the right quantity proved impractical and ate into the theoretical profits. "The valuable constituents of sewage are like the gold in the sand of the Rhine," one chemist said. "Its aggregate value must be immense, but no company has yet succeeded in raising the treasure." By the final quarter of the nineteenth century, the vision of making loads of money on fertilizer from sewage all but disappeared.

The development of the activated sludge process in 1914 led to the revival of a more modest version of that dream, writes environmental historian Daniel Schneider in his book, *Hybrid Nature*. Two years earlier, in search of solutions for sewage treatment, chemist and bacteriologist Gilbert Fowler of the University of Manchester in England traveled to a public health research laboratory in Lawrence, Massachusetts (about thirty miles north of Boston), where he saw experiments that showed that bubbling air through sewage in the presence of microbes and other life-forms helped to quickly separate it into a clear liquid and a sludge. Back in Manchester, he and his colleagues discovered that they could turn this into a continuous process by capturing some of the sludge and feeding it back into the process, thereby building up a healthy, active (or "activated") population of naturally occurring microbes, much like how a baker cultivates a starter for sourdough bread. The clear liquid effluent from this microbe-driven treatment could be more safely discharged to rivers or other bodies of water than raw sewage. The solids—mostly leftover microbial cells—it turned out, had a high enough nitrogen content to make it promising as a kind of fertilizer. Milwaukee was the first large city to adopt this new technology. While using Lake Michigan as both a waste receptacle for its sewage and a source of drinking water, it had suffered a typhoid outbreak in 1916, which sickened thousands and killed sixty. Its officials justified the high costs of a new treatment plant by arguing that it would make back some of the money by selling the residue. All they needed to do was make it dry enough to transport and apply, which they did with a three-stage process that included settling, pressing, and heat drying.

Commercial production began in 1926, under the trade name Milorganite, a portmanteau of *Milwaukee organic nitrogen. Organic*, in this context, means that the nitrogen originates in living material. In the soil, microorganisms slowly convert it into inorganic versions that plants can use. Synthetic nitrogen, in contrast, is water-soluble, so an ill-timed rain can quickly wash it through to the water table, skipping plant roots, wasting money, and potentially polluting a drinking-water source or contributing to dead zones in water bodies. Because of its origin in sewage, Milorganite also contains a variety of micronutrients that plants need, as well as organic matter, which improves the structure of the soil, including its ability to hold water.

Landscapers realized that turf grass, in particular, thrived on

Milorganite's particular chemical composition, so the product developed a niche: golf courses. The love affair between Milorganite and turf managers continued for decades. You can see it in the popular 1980 golf comedy *Caddyshack*, in which the oddball assistant greenskeeper, played by Bill Murray, plots the murder of a ruinous gopher in a shed full of green-and-orange bags of Milorganite. Despite Milorganite's popularity, however, it was never easy money: selling the stuff for a decent price required sustained marketing and continuous innovation, not to mention keeping up with developments in wastewater treatment processes as well as state and federal regulations. Other municipal treatment plants at first found it easier to just dispose of the residue sludge in some nearby water—that is, until 1972, when the Clean Water Act regulated disposal in waterways and coastal areas, and 1988, when the Ocean Dumping Ban Act did the same for the whole ocean.

Ultimately, to make the sludge product more palatable to consumers, a wastewater treatment industry organization, the Water Environment Federation, decided it needed a pleasanter name than *wastewater solids* or just *sludge*, as it had been called until then. In 1991, following a call for suggestions that garnered three hundred entries, they dubbed the substance "biosolids." The market ate it up. In 1993, the EPA developed new rules for biosolids—including classifying them according to levels of pathogens, metals, odors, and other factors—writing that it "believes that biosolids are an important resource that can and should be safely used." Class B biosolids don't require as much treatment, and their use is restricted. But Class A and Class A EQ (Exceptional Quality, meaning that it exceeds Class A requirements), which is what Pawlowski's factory makes in Quincy, are "pretty much unrestricted use, which would mean you could use it in your home vegetable garden."

If we could fully recover all the nutrients from wastewater, according to one estimate, we could offset about 13 percent of the global fertilizer nutrient demand, with a revenue of almost $14 billion. Likewise, if we could convert it to biogas, it could provide electricity to 158 million households annually. In fact, wastewater itself contains more energy than what's required to treat it using conventional methods, but making fertilizer and energy during treatment are far from universal practices.

And there's even more that the sludge can become. In a Dutch town called Epe, operators installed a modified conventional treatment process known as Nereda, which uses less energy and requires less space. Those are good enough reasons, but it also turns out that the sludge granules from the Nereda process contain bacteria that can form a useful biopolymer. Engineers have dubbed it Kaumera Nereda Gum, after the Maori word for *chameleon*, since it is so versatile—it both absorbs and repels water. I have a sampler tube of eight multicolored Nereda gumballs sitting on my desk, looking just like the ones that used to come from ubiquitous glass-globed machines. Although customers for the weird stuff still need to be courted, the inventors claim that it could replace a substance called alginate, derived from kelp, which is used in paper production, plaster molds for medicine, and food stabilizers. Researchers have also incorporated it into a biodegradable concrete coating that could replace petroleum-based coatings. The designer Nienke Hoogvliet found that it helped textiles absorb dyes better, so less water was needed, using it to make a stunning blue-and-orange kimono, which she exhibited at the Dutch Design Week in 2018.

Using poop to produce food and other products may sound disgusting, but if our planet is to be self-sustaining—and if we are going to send self-sustaining missions to new frontiers—we will need to take a rational approach to the risks and rewards of reusing the rich resources in our waste. We will have to think like astronaut Mark Watney, the hero of Andy Weir's novel *The Martian*, which is also a movie featuring real-life toilet enthusiast Matt Damon. Stranded in a Martian research station without enough food to eat, Watney fertilizes Martian soil with his crewmates' vacuum-dried feces so that he can grow potatoes, spreading it all around his living space. It's disgusting, but, more important, it keeps him alive. "The worse it smells," Watney writes in his journal, "the better things are going."

Gone to Pot

In the 1980s, Joe Jenkins moved to a remote property in northwestern Pennsylvania. With no electricity to pump water, he turned to composting for his sanitation needs. He compiled what he learned in the self-published *Humanure Handbook* in 1995, and, to his surprise, it sold thousands of copies and got translated worldwide. In 2019, he put out a fourth edition, which retails for $25. One of his commandments: don't call his loos compost*ing* toilets—the toilets don't compost; you do.

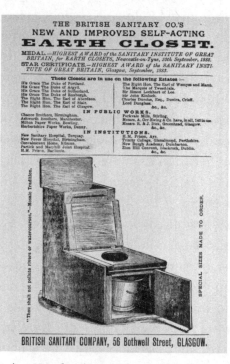

An 1884 advertisement for an earth closet.

Jenkins is an intellectual heir of Reverend Henry Moule, vicar of Fordington, a village in the southwest of England. While other Victorian sanitary reformers pushed the water closet, Moule designed several "earth closets," earning his first patent in 1860. Essentially, an earth closet was a wooden chair with a hole in the seat and an iron bucket underneath. Nearby—often in the backrest—was a hopper filled with soil, with which to cover the waste after every use. Based on his home experiments, Moule knew that mixing feces and earth together, preferably without urine, and then drying it in an open shed would transform the feces into useful compost (much as the organization SOIL does in Haiti). Today, we know that microorganisms—mainly bacteria and fungi—do the work of decomposition. They need moisture and oxygen and give off carbon dioxide and heat. That heat kills off pathogens, leaving behind a humus that's nothing like poop.

Moule imagined these earth closets everywhere, but because composting in this way required labor and organization, they worked best in managed communities, where a dedicated employee could make and distribute that cover soil, maintain the closets, educate the users, collect the buckets, and keep everything dry. Some of Moule's devices were even semiautomatic: a spring system in the seat caused soil to release every time someone stood up. If the system broke down, however, it would start to smell—or worse. In Wakefield Prison, a typhoid outbreak in 1874–1875 led investigators to find that some of the hundreds of closets in use there were "most disgustingly dirty through misuse and neglect" and probably contributed to the deaths of eighteen inmates.

Still, the earth closet continued to appeal to a limited set of rural folk, who had plenty of soil and not much running water and preferred it to the hole-in-the-ground privy because of the ease of recovering nutrients. Some persisted in the English countryside until the 1950s and 1960s, until flush toilets and septic tanks prevailed. But the idea of the off-grid compost toilet didn't totally die—every few years or so, someone declares that the trend is back, driven by environmental and humanitarian concerns or just a desire to be self-sufficient. "Humanure: The End of Sewage as We Know It?" asked the *Guardian* in 2009. In 2014, it published "Composting Toilets: A Growing Movement in Green Disposal." Then, in 2019, "The No-Flush Movement: The Unexpected Rise of the Composting Toilet." All along, these toilets have been living a rich underground life, spawning ever-new approaches and recruiting new adherents.

One researcher, after experimenting with a variety of toilets, happened to learn about an ancient Amazonian creation, *terra preta*—"black earth" in Portuguese. Archaeologists have found this highly fertile man-made soil across the Amazon basin and have concluded that indigenous people generated it by combining charcoal, bone, manure, and other organic materials. Astounded by its longevity, Ralf Otterpohl, of the Technical University of Hamburg, came up with what he calls *terra preta* sanitation, which differs somewhat from Moule's earth closet. Instead of soil, users immediately cover their feces (again, no urine) with a charcoal mix, which both removes the odor and provides stable organic matter. Then they add a few dashes of a microbial mix that includes lactobacilli, the lactic acid bacteria that also can produce fermented products such as sauerkraut, kimchee, and yogurt. (Otterpohl recommends taking the bacteria directly from sauerkraut liquid, which is, of course, abundant in Germany.) Then the container gets sealed up (and replaced with an empty one), and the bacteria get to work in an anaerobic environment for at least a month. This process raises the acidity of the mix, killing off pathogens. At the worst, he says, you get "not this nasty fecal smell, but something sour." After this step, the mix goes into a compost heap where, for three to six months, worms eat the decomposing material and poop out vermicast, which Otterpohl and his colleagues claim is a lot like actual *terra preta*.

Pit latrines, common in low-income regions of the world, can also produce compost. In so-called *twin-pit latrines*, a toilet—which may flush, should the user want to, since the system can handle a little water—sits over one of two pits, which are side by side. After one

pit fills up, the toilet alternates to the second pit, while the first pit
"rests" and the sludge in it decomposes. After two years, the contents
can be shoveled out and used as compost and the pit used again.
Sociologist Bindeshwar Pathak, the charismatic founder of the Sulabh
International Social Service Organisation in India, came up with these
and spread them throughout throughout the country. And then there's
the Arborloo; after the unlined hole in the ground is full, the user
moves the toilet structure to another spot and plants a tree where
it used to be, allowing the nutrients to return to food in the form of
banana, papaya, or guava.

After any of these composting processes is complete, all the original
fecal material should be gone, eaten by the microbes, and the pathogens
killed off. In short, these composts *should* be as safe as any other compost from, say, food waste. But many experts recommend exercising
caution when using it with food crops. As for Jenkins, "I have always
used humanure as a feedstock for the compost," he writes. "I have
also used all the finished compost to grow my food (some has gone to
houseplants), and I have raised a healthy family on my garden produce."

One new technology for recycling toilet waste back to the earth originated not with toilets but with dog droppings. In 2007, Israeli entrepreneur Oded Halperin was talking on the phone while walking his
pug in Tel Aviv and didn't notice when the dog pooped. Unfortunately
for him—but perhaps fortunately for the world—a police officer saw
and fined him. Frustrated, he decided to find a better way to clean up
dog poop than picking it up with a bag-covered hand. He contacted a
nanobiotechnologist friend with the same first name, Oded Shoseyov
of Hebrew University in Jerusalem, a "mad scientist," as described by
their American business partner Aaron Tartakovsky. Could he come
up with something new? After tinkering around with poop and many
different chemicals, Shoseyov homed in on one, potassium permanganate, that, when blended with poop at high speeds, neutralized the
odor, evaporated out the moisture, and killed off pathogens. Potassium

permanganate is a common substance used for purposes including water treatment, odor control, and disinfection. It breaks down into potassium and manganese, which naturally occur in soils, so the resulting dry substance, which has the consistency of ash, is suitable as a biosolids-like soil additive.

The first result of this invention, which briefly went viral in 2011, was a high-tech pooper scooper called the AshPooPie. It never made it to market, but the founders also decided to apply the technology to human waste. One spin-off, the San Francisco–based company Epic CleanTec, aimed to develop the technology for buildings, as an example of distributed infrastructure. Getting poop to experiment on wasn't easy, says CEO Tartakovsky: They started their work in a garage, sourcing dog poop from local kennels. Then they moved to Stanford's Codiga Resource Recovery Center, a facility dedicated to testing new water and energy resource recovery systems, which allowed them to collect sewage from campus, buying waste from the university by the pound. Next, they worked with a pumper truck company that collected septic sludge from around the Bay Area, including from the Facebook campus. Later, they established a pilot system in a thirty-five-floor luxury residential tower in San Francisco, transporting the solids to a hub in a former Honda dealership, where they treated them with the new technology and used the output to grow a garden. Now, with a large national real-estate developer, they're planning a full commercial installation in a fifty-five-floor mixed-use high-rise, also in San Francisco.

One of the main benefits to the in-building system is that it captures the solids within minutes of flushing, Tartakovsky says, so turds and toilet paper don't have time to break down and dissolve into the water, making it easier to filter them out again. Toilet paper is welcome, Tartakovsky says, since it adds to the organic matter in the product, which they hope to package and sell to local consumers, including home gardeners, under their own brand name. Treatment with the technology takes just about a half an hour per batch—a far cry from the months it can take to compost the traditional way. And they can clean the remaining water with membranes and other advanced treatments for reuse in toilet flushing, landscape watering, and air-conditioning. Since specific regulations don't exist for everything that they're doing, they've decided to "self-impose" biosolids standards and have kept an open line to the EPA, as well as public health and building regulators. This technology is the best way, Tartakovsky believes, to get the advantages of composting and water reuse without asking people to make big lifestyle changes. "The vast majority of people, when they go to the toilet, they want to see water in the bowl," he says.

Today, many people who willingly forego the flush and adopt one of the modern versions of the earth closet do so because they need one for a rural retreat, boat, or recreational vehicle, or want a second or third toilet at home for a workshop or attic. While some follow Jenkins's instructions and construct a simple wooden box, many others purchase commercial models, called dry toilets. A popular one from Nature's Head looks something like a standard pedestal toilet and costs nearly a thousand dollars, while ones that aerate or heat the contents can cost a good deal more. But while the growing availability of these toilets might inspire some trend articles, it doesn't indicate that all of Jenkins's methods are becoming mainstream. Many owners of compost toilets just drop the contents in the nearest dumpster, while only a brave few boldly take on the challenges—and risks—of creating humanure.

Super Fly

In the 1930s, health officials were trying to introduce a new kind of pit latrine to a region in Louisiana with poor sanitation when unwanted visitors invaded the project: black soldier flies. As the larvae hatched in the toilet waste and crawled out of the pits, they attracted chickens, who destroyed the toilet in their rush to feast. Appalled, the researchers poured gasoline down the toilet to kill the larvae.

Maybe they shouldn't have. Today, sanitation engineers are farmers in a sense: they must carefully tend to the microscopic organisms that they need in their systems and fend off the ones that cause trouble.

But, for a long time, they've had a limited notion of which organisms belong and which do not. Innovators are now reconsidering these old ideas and inviting a new set of poop-eating creatures to the toilet feast.

Among the candidates are tiny aquatic plant-like microorganisms called microalgae, which might be able to harvest the nutrients from toilet waste while leaving more of the pollutants behind. In environmental engineer Tania Fernandes's lab at the Netherlands Institute of Ecology (NIOO-KNAW), tubes that looks like lava lamps glow different shades of green. In them, microalgae consume not only the major nutrients in wastewater, which are nitrogen, phosphorus, and potassium, but also the minor ones, such as zinc and cobalt. This broad spectrum means the microalgae can serve as a fertilizer that's more complete than conventional ones, Fernandes says. But microalgae grown on wastewater might also have other potential uses, such as dyes for textiles. Since they're powered by photosynthesis, they are a good fit for warm, sunny places such as India—and less so for cool, gray ones such as the Netherlands.

Another option is to grow protein-rich microbes on the gases that come off a treatment process. In the 1960s, NASA funded research into whether this would be a good solution for long-term space journeys, when astronauts couldn't get regular deliveries. Scientists showed that a bacterium, called *Hydrogenomonas eutropha* (since renamed *Ralstonia eutropha*), could thrive on the nitrogen and hydrogen by-products of a zero-gravity wastewater treatment system. When scientists raised lab rats on a diet of the bacteria, the rodents grew well. But when they tried feeding the stuff to primates, poor results suggested that it might contain a possible toxin. More recently, researchers from the Pennsylvania State University published a similar concept. The first step would use anaerobic digestion to break down pee and poop into salts and methane gas. Then the methane would feed what the researchers described as "goo." In their prototype, they grew *Methylococcus capsulatus*, which is used as animal feed today and might also be suitable as human food, though it hasn't been tested. Their reactor made a lump of it that was a nutritious 52 percent protein and 36 percent fats.

For earthbound wastewater treatment plants, bioengineer Willy Verstraete, an emeritus professor of Ghent University in Belgium, has put forward a similar concept, "upcycling" what he calls the "used nitrogen" in our pee and poop with the help of microbes. Livestock or people could eat the resulting goo, helping to address the growing demand for protein as more of the world shifts from plant-based to meat-based diets, leading to overfishing and deforestation. He believes that, since the microorganisms consume only the gases from the wastewater treatment process and not any of the actual fecal matter, the "yuck factor" would be lower. But how would it taste? The microorganisms are unlikely to have a yummy texture or taste on their own, he says. Maybe, for the sake of survival, astronauts would be willing to gulp down an unappetizing Vegemite-like schmear on a bagel, but most people on Earth wouldn't accept it as a substitute for, say, deli meat. To make it more palatable, Verstraete says, "we need food scientists."

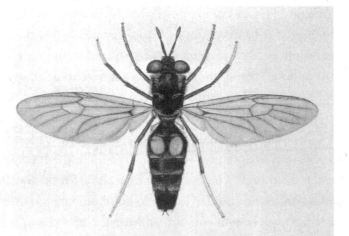

An illustration of a black soldier fly, *Hermetia illucens*.

And then there are the flies. Starting in the 1990s, researchers figured out that the black soldier fly (*Hermetia illucens*) has a combination of traits that makes it ideal as a biological waste processor. These large, slow flies are unusual in that they don't eat any food in their adult, winged stage, so they do not pester humans or spread diseases like other flies that flit from feces to food. Instead, in their younger, worm-like larval stage, they eat ravenously, growing into a plump morsel for other animals, especially fish and poultry. On top of that, the species can live almost anywhere. Native to the warmer regions of the Americas, the fly spread around the world's tropical and temperate regions starting in the 1940s due to the rise of international transportation. By the 1960s, it had made itself at home along the Mediterranean coasts of Spain, France, and Italy, and it's even been spotted as far north as Germany and New Hampshire. Scientists have reared the pale, segmented larvae on a huge variety of wastes, including rotting fruits and vegetables, coffee bean pulp, pig and poultry manure, and, of course, feces.

Today, several companies are trying to make black soldier fly–powered factories that process various kinds of organic waste. One of the most prominent is AgriProtein, named in 2018 one of *Time* magazine's fifty "genius companies," which is developing black soldier fly technology for food waste. In Cape Town, I take a tour of their demonstration facility, which has just completed an upgrade, allowing it to farm billions (yes, billions) of flies in different life stages, as well as accept more than forty tons of food waste per day. In a borrowed lab coat, I follow my guide, Rozane Badenhorst, the global biology manager, through dark, warm, and humid rooms, where buzzing flies mate in rows and rows of neat white mesh cages under blue lights that put them in the mood. The males put on a show for the females, and then each couple grab ahold of each other and plummet to the ground. "If you listen closely, you can almost hear the flies dropping to the floor" with a "*tuk, tuk, tuk,*" Badenhorst says. "It's actually therapeutic sometimes." After laying her eggs, the female dies.

Workers collect the tiny eggs and take them next door to incubators to hatch into larvae no bigger than grains of sand. Transferred to bins, these larvae grow for four days on a specially formulated "nursery feed," until they are about the size of rice grains, and then for eleven days on processed food waste. When the pale, segmented critters are about an inch long and almost ready to start morphing into flies, it's time to harvest them: they're killed with steam and ground into two proprietary products. One is a protein-rich powder, which the company can sell to farmers and aquaculture companies as a component of animal feed. The other is an oil, which can be mixed into feed and also has applications in cosmetics and industrial processes. The uneaten waste—mostly the larvae's poop—can be used as a compost-like fertilizer. The feed is more environmentally sustainable than some other options, such as fish meal, which contributes to overfishing, and soybeans, which are linked to deforestation and unsustainable agriculture. Fast-food giant McDonald's has declared an interest in increasing its use of insects as chicken feed.

With a high-speed video camera, researchers from the Georgia Institute of Technology in Atlanta have examined just how black soldier fly larvae manage to eat so quickly, going from hatchling to plump morsel in as little as two weeks. "We saw that this larva has these two teeth that are called maxillae, and the larva uses it to shear pieces of food like this," says mechanical engineering graduate student Olga Shishkov, moving her index fingers alternately up and down, as if two-finger typing. "But it's not a very powerful mouth," so it makes sense that they prefer soft foods. Then, using time-lapse video with cameras mounted above and below a fish tank to analyze the larvae's collective motion, Shishkov and her colleagues found that the eating larvae flow like a self-pumping fountain. Each larva wriggles toward the center, chows down, bursts out of the top, and falls down the pile. For every satiated larva, there's a hungry one just behind it.

Black soldier flies can feed on feces instead of food waste. AgriProtein's sister company, The BioCycle, has been trying to make it work in Durban. Following a cholera outbreak in 2000, the eThekwini Municipality installed about eighty thousand dry toilets on its rural outskirts. At first, households buried the sludge from the new dry toilets on their properties. But, years later, the government realized that handling the sludge put people at more risk than originally thought; plus, people didn't like to do it. The municipality needed to start collecting the poop. With support from the Gates Foundation, eThekwini and The BioCycle designed and built a large facility to process more than twenty tons of fecal sludge per day.

Since the facility opened in 2017, however, the project has had problems. For one, it had to confront trash in the sludge. That became a sticking point among the company, the city, and the contractor hired to empty and transport the waste. And while AgriProtein spends a lot of time optimizing the feed and the climate for the larvae in Cape Town, the Durban BioCycle facility was supposed to be low-tech—just an open shed—so that the concept could transfer to low-income contexts. But the larvae weren't thriving. The company brought in mobile climate-controlled units, scaled back to a research and development operation, and started burying most of the delivered toilet waste nearby.

On top of the technical problems, there is also the question of who will buy the products. After all, potential customers are likely to be wary of the fecal origins until it is proven absolutely safe. And, in some countries, preexisting legislation prohibits the sale of the insects as feed for at least some animals. Still, The BioCycle would prefer to turn them into other types of high-value products. The maggot-derived oil could potentially replace palm oil, which is linked to massive deforestation in Indonesia, in industrial uses. And scientists are figuring out how to extract chitin, a fibrous substance found in arthropod exoskeletons that can be used in medicine as scaffolding for organs. Some are even developing uses for chitin in the fecal sludge treatment process. Today, most commercially available chitin gets extracted from shellfish waste through a polluting and expensive process.

Looking at the Swedish market, Cecilia Lalander of the Swedish University of Agricultural Sciences in Uppsala estimates that black soldier fly larvae are among the highest-value products that could be made from feces; the same is likely to be true in low- and middle-income contexts. In Kenya, the main sources of protein and fat for animal feed are troubled, says Frederick Wangombe, a veterinarian and animal nutritionist who mixes feed for farms. "Fish meal is seasonal, and prone to adulteration. Soybean meal has also become seasonal in the last couple of years." The larvae, on the other hand, have the potential to be more consistent in both quality and year-round availability. He currently sources some from an enterprise called Sanergy, founded by three Massachusetts Institute of Technology alumni, which services more than three thousand of their own Fresh Life brand of toilets in the informal settlements of Nairobi, collecting toilet waste and composting it with agricultural, market, and kitchen waste. Recently, Sanergy has developed its own colony of black soldier flies and is now expanding its operations to accept nearly eighty thousand tons of toilet and other organic waste per month to turn into insect-based feed, compost, and other products. The larvae make "an excellent meal for poultry and pigs," Wangombe says, and he's eager to get more. As for the ick factor? The farmers he works with don't mind the origin of the feed, he says, so long as their animals thrive.

Poo Power

American engineer Emily Woods remembers the moment well: she and her colleagues were in a Kenyan friend's backyard in Karagita, a small town on the edge of Lake Naivasha. They were impressed with themselves, having figured out how to desiccate feces in a solar concentrator that looked like a large, reflective satellite dish. "We built this parabolic mirror to focus solar heat on a black metal bucket filled with poop," she explains to me. But when the friend showed it off to her neighbors, they had another idea for the device, which amounted to a solar stove: "Can we cook on this?" they asked. It turned out that locals were spending large portions of their income on firewood and charcoal to make meals. Fuel, Woods realized, was a far more pressing concern for many Kenyans than toilets.

That was 2013. Soon after, Woods and fellow engineer Andrew

Foote co-founded the social enterprise Sanivation to put these two ideas together, not by selling solar stoves but by producing poop briquettes for people's existing stoves. In Kenya, charcoal usually gets made from wood in inefficient traditional earth kilns, contributing to deforestation and climate change. But Sanivation collects poop in containers, treats it, and presses it with other kinds of waste, such as charcoal dust and rose waste, to make "poop fuel"—soon, more than a thousand tons per month. Though cooking with poop may sound unappetizing, the Sanivation briquettes have proven popular with home cooks since they burn cleaner, last longer, cost less, and save the country's trees.

Today's innovators are far from the first to try to unlock the energy in poop: in the biblical Book of Ezekiel, which is written in the voice of a priest living in Babylon in the early sixth century BCE, God tells Ezekiel to cook a barley cake over human dung. Up-and-coming technologies range from the low-tech, such as Sanivation's briquettes, to the futuristic-sounding, such as plasma gasification, which is essentially microwaving. For some advanced treatment plants, combining anaerobic digestion (which produces biogas) with energy-saving improvements has helped make them more self-sufficient—a few have even managed to achieve energy neutrality. But many of the smaller, more cash-strapped treatment plants in the United States don't take advantage of anaerobic digestion because they can't afford the costs of installing and running the equipment needed to either make the gas or upgrade it to a form they can use—startlingly, many of these plants flare off the biogas that they make.

Many new concepts combine energy, fertilizer, and water reuse, like the one under development by environmental engineer Daniel Yeh's team at the University of South Florida. It had a moment of fame in 2018 when it was featured on *The Daily Show* with Trevor Noah. The system combines anaerobic digestion with a high-tech membrane filter, which cleans the water that comes out to a very high quality. He calls the system the NEWgenerator for the nutrients, energy, and water that it produces. About a third of the size of a standard twenty-foot shipping container, "it's a black box, literally a box, and in will come sewage and out will come these beneficial products." He has even started working with NASA to see if the concept could be adapted for future missions to Mars. Urban slums and Mars colonies share in common that they are "resource-limited systems," he says, though the former tend to have little money while the latter would have budgets that are literally astronomical.

About a century ago, scientists proposed another tantalizing idea. A combination of naturally occurring pressure and heat, over millennia, once turned ancient organic matter into today's fossil fuels. Could machinery reproduce that process, much faster, to turn today's organic matter—which includes solids from sewage, of course—into petroleum?

The idea, called hydrothermal processing, remained little more than a hypothesis until the 1970s. That decade, an embargo by the Arab members of the Organization of Petroleum Exporting Countries drove up fuel prices, creating worldwide interest in renewable fuels. At about that same time, the Pacific Northwest National Laboratory (PNNL) in Richland, Washington, which had been founded to develop atomic weapons, needed a new purpose. Renewable energy fit the bill, and hydrothermal processing seemed like a worthy target.

Progress was slow, but one of the advantages of national laborato ries is that they can grind away at long-term projects that would put private companies out of business. "In the early days, most [of the research] had to be done with experimentation and trial and error," says James Oyler, president of Genifuel Corporation. "There weren't massive databases or lots of computers to do the analysis." Over the next forty years, he says, even as oil prices fell again, PNNL scientists gathered the data they needed to refine a process that can essentially pressure-cook organic material. The water in the material sits just below its supercritical point, where it remains liquid but gains special properties, causing chemical reactions that break down complex molecules in minutes instead of millions of years. The outputs are crude bio-oil, which a standard refinery can then transform into any kind of liquid fuel, from diesel to jet fuel, as well as methane gas, another useful fuel.

Just as you don't want to put a wet log on a campfire, most energy-producing processes don't do well with wet materials. But hydrothermal processing is the opposite: the water is key to how it works. Over time PNNL has tested the process on "everything from fish heads to marigold plants to chicken manure to algae," Oyler says. "You name it, probably something has been tested with it."

Still, nobody knew quite what to do with the technology. Oyler first started Genifuel in Salt Lake City, Utah, in 2006, after decades as an executive at technology, engineering, and aerospace companies. His first goal was to use a different process to make diesel fuel from farmed algae, not sewage. When that didn't pan out, he heard about PNNL's hydrothermal-processing program and got in touch. They agreed to throw some of his algae into the national lab's machine. It "worked really well," to his delight, readily turning the algae into fuel. But it wasn't commercially viable because it would cost more to grow the algae and process it than he could sell the fuel for.

But Oyler, whose name makes him sound destined for this work, came across another idea. He could trade out the algae for another wet substance: sewage sludge. He would not have to invest in producing the raw material, since it already exists in abundance. And wastewater treatment plants were spending money—sometimes more than half of their operating costs—to get rid of their residual sludge, so all he would have to do was undercut that price. It would take some more work to get it right, but Oyler's a calm, methodical type, and PNNL already had put forty years into it—a few more would hardly matter.

Today, after building eight successful prototypes of increasing size, Genifuel is working with Metro Vancouver in Canada to build a demonstration facility that will capture the sewage from thirty thousand people and turn it into five barrels of oil per day. Using temperatures of about 680 degrees Fahrenheit (not quite as hot as a pizza oven) and pressures of about 3,000 pounds per square inch (similar to a heavy-duty pressure washer), the technology is calibrated to turn the carbon-based matter in sewage into bio-crude and methane gas. The fuel product contains much more energy than the system needs to run, Oyler says.

In addition, the process spits out all of the water that's put into it, "except, when it comes out, it's sterile and clean," Oyler says. The water contains potassium and nitrogen, as well as trace metals such as iron, copper, and zinc, all of which plants need to grow, so it "can actually be used as a liquid fertilizer." (The plant pulls phosphorus out of the system at an earlier stage because it's a nuisance, but that, too, can be made into fertilizer.)

The business case is also good. Anaerobic digesters take about a month to process a batch of solids and reduce them by half, which means that the utilities still have a lot of solids to get rid of. A Genifuel hydrothermal-processing plant, on the other hand, runs in a continuous cycle that takes forty-five minutes and consumes all of the organic solids. "Now they don't have to worry about what to do with it, where to put it, how much it costs to haul it away, how far they have to carry it—none of that—and they get oil and gas." And acquiring the technology isn't even a terribly onerous up-front investment, Oyler says, since its cost would be comparable to that of a new anaerobic digester and could be added whenever existing equipment comes up for replacement or major repairs.

Research from PNNL estimates that if hydrothermal processing were applied to all of the sewage treated in the United States, it would generate about 27 million barrels of crude bio-oil per year, which, once refined, could meet more than 4 percent of the typical U.S. demand for kerosene-type jet fuel, the most common type of aviation fuel. If food-processing waste, cow and pig manures, and other wet wastes were also added to the sewage sludge, Genifuel-type facilities could meet nearly a quarter of that jet-fuel demand. At least for now, though, the technology is probably not appropriate for low-resource contexts, so it won't put sanipreneurs like Sanivation out of business anytime soon.

To a lot of people, it sounds too good to be true, Oyler says. But Genifuel's a remarkably open book, at least scientifically. "By working with a national laboratory, everything we do is public." The demonstration plant will show whether the technology can work on a large scale, day in and day out, without fail, as utilities need it to. In the end, what this technology promises is to take away one of wastewater operators' greatest worries, sludge, and—*poof!*—turn it into something that makes cars go and airplanes fly.

Modern Manure

So that I can see for myself how sludge becomes Class A biosolids, Carl Pawlowski, the boater-turned-fertilizer-supervisor, arranges to show me around the factory on a Friday afternoon when I am in town. I take the T—Bostonian for "subway"—south from the city center to Quincy, where I hop a bus. After walking several long blocks past modest single-family homes, I cross railroad tracks lined with coal-black tanker cars, only to meet a giant, faceless brick warehouse. With my rolling luggage and massive backpack, I feel like a lost laden ant trying to find my way back to the colony. Finally, with the help of a few lunching employees of the Boston Aquarium—which has a building in the industrial park—I reach the front door of the plant, labeled "New England Fertilizer Co." That's the company that holds the contract with the water authority to run the operation. Pawlowski meets me inside. Clean-cut and comfortable in a short-sleeved button-down shirt and khakis, with close-cropped salt-and-pepper hair, he speaks intensely, often moving his reading glasses between his nose, the top of his head, and his front pocket.

I come by land, but the raw materials for the fertilizer come by sea—under the harbor, in fact. Thanks to the $4.7 billion program that cleaned up Boston Harbor (of which $87 million went into the fertilizer factory, opened in 1991), if you poop in greater Boston today, your contribution first flows through sewers to Deer Island, one of the world's largest wastewater treatment plants, just east of Logan International Airport. The solid residue from the activated sludge process flows into six giant cement eggs as tall as downtown office buildings, where anaerobic digestion takes place, releasing biogas that then powers and heats the plant. The sludge that comes out of those eggs is three generations removed from your poop: it's microbes that ate the microbes that ate the poop. And that's what flows southward through a different underwater pipe seven miles across the bay to the factory in Quincy, on the banks of the Weymouth Fore River, where four 1-million-gallon storage tanks await it. I see the solids pipe on both ends: appropriately, it's poop brown and labeled "DIGESTED SLUDGE" with an arrow pointing in the direction of flow.

Once in Quincy, the main problem with the sludge is that it's still 98 percent liquids. The plant must get as much of those remaining liquids out of it as possible to stop any biological activity by the microbes, as well as make the product lighter so that transport is cheaper. And it has to be done quickly, since Pawlowski has just a small corner of Quincy, and a whole lot of sludge to process: 1.2 million gallons per day.

For that, they use machines. The first is a centrifuge, which, although it is not running during my hard-hat tour of the factory, emits a pungent ammonia smell—one of the few strong odors I notice on the otherwise earthy-smelling plant floor. A turquoise cylinder on the outside, inside it's shaped like a large old-fashioned glass Coca-Cola bottle placed on its side. It spins rapidly, forcing the solids to separate from the liquids with the denser solids sticking to the inner wall of the centrifuge and the liquids resting on top. Rotating screw blades scrape the solids—which are like "wet earth"—up and out of the mouth of the bottle, while the liquids flow with gravity in the opposite direction, discharging through a separate outlet. After that, the solids get mixed with tiny "seed pellets"; they glom on like fresh snow to snowballs, forming the bigger pellets that fit into farmers' and landscapers' spreaders. Then the pellets drop into a dryer that's "just like your clothes dryer," Pawlowski says, which heats the pellets to about two hundred degrees Fahrenheit, so that no pathogens survive and they come out at about "ninety-six, ninety-seven percent solids," he says.

After they're dry, the pellets go onto a series of screw conveyors, which sort them for size. Pieces that are too small get recycled into seed pellets for the pelletizing process; those that are too big get crushed into seed pellets. Pieces of just the right size to fit into fertilizer-spreading machines—about one to three millimeters in diameter—continue on to a pellet cooler and then get pushed through pneumatic tubes to storage silos.

These storage silos are the part of the operation that Pawlowski worries about most. They don't look too imposing: white cylinders standing on end, just a few stories tall, with two blue stripes wrapping around the top like a hospital baby blanket. But you may have heard stories about grain silos exploding—and, well, that could happen to the plant's fertilizer silos, too, if there's microbial activity in the pellets and too much built-up dust. The metabolizing microbes could create so much heat that they could start a fire, and the dust could combust.

Each silo has a grid of thermometers that monitors the temperature throughout. If they notice a "hot spot," the silos trickle nitrogen down to displace the oxygen that the microbes need. "We want the microbes to asphyxiate," Pawlowski says. Fighting this hazard is kind of like playing a never-ending video game against the microbes. In the control room, he points to one 118-degree hot spot on a dim old Dell computer monitor: it's getting the trickle. And there's another way that Pawlowski manages this risk: "You keep product moving" by pricing it right, even in winter when there's less demand. And you sell it locally. That way heat and combustible dust don't have time to build up during storage and transit.

*   *   *

A big part of Pawlowski's job is reassuring the public of the safety of the finished product. A few years ago, a farmer in Connecticut got a bulk delivery of his fertilizer. Instead of spreading it out, as he should have done, he left it in a heap. It rained, and the rainwater activated the microorganisms that had colonized the pile. "The inside started to smolder, and that smells really bad," Pawlowski says. The neighbors noticed and called the authorities, who came out to investigate. "Don't worry, folks; it's just biosolids," the police told the neighbors. Confused and concerned, the neighbors looked up *biosolids* online and put together that the stinking pile originated as sewage. Soon, one of them made a page on Facebook to alert the community. It went viral.

Pawlowski maintains an outreach program, including a "little rinky-dink bagging operation" that puts a small percentage of the pellets into attractively designed bags labeled "Bay State Fertilizer" (the bulk product is called New England Fertilizer). While we're talking, a sturdy landscaper walks into the factory's offices to pay for his purchase. It's guys like that whom he can rely on to vouch for the product when concerns arise, Pawlowski says. "I could fill a room with a hundred people who would all stand up and say, 'I've been using this for ten, twenty years, it's perfectly safe, it's perfectly normal, I've never had an issue with it.'"

Around the world, biosolids are often controversial. Responding to public sentiment—sometimes stoked by a small group of fervent activists—several major food processing companies have declared that they won't make products from produce grown with the help of biosolids. In 2014, Whole Foods announced it would stop selling biosolids-grown items, even though little to none of their stock was grown in biosolids at the time anyway. And quite a few U.S. counties and the entire country of Switzerland have banned biosolids use altogether.

It's not just that the public finds poop gross, though that is likely part of it in some cases—it's also about balancing the pros and cons of the options for dealing with sewage sludge. While biosolids have advantages, new attention to some of the disadvantages may cause the balance to shift away from them. Writing in the journal *Environmental Science & Technology* in 2015, engineers Jordan Peccia of Yale University and Paul Westerhoff of Arizona State University called today's widespread use of biosolids on soils "good intentions, wrong approach," asking: "How did sludge reuse and disposal policy get to the point where municipalities pay for the majority of a wastewater treatment plant's energy, hazardous chemicals, and pathogen content to become embedded in biosolids, which are then dispersed into the environment?"

Class A biosolids must meet strict limits for certain pathogens and pollutants. But, as with the liquid output of conventional sewage treatment (which, in Boston, discharges through a 9.5-mile-long tunnel into the bay), trouble still lies with increasing numbers of unregulated household and industrial substances. About two years after my visit to Pawlowski, a fresh controversy raised its head with a report from the *Boston Globe* that revealed that the biosolids from his plant, like many plants around the country, contain what some consider to be concerning levels of PFAS, the abbreviation for per- and poly-fluoro-alkyl substances. PFAS are a class of thousands of water-, grease-, and dirt-repellent chemicals that have been made since the 1940s. They break down so slowly, over geological time, that they're sometimes called forever chemicals. Although they are useful additions to many products such as nonstick pans and firefighting foam, they have also been linked to low infant birth weights, kidney and testicular cancer, and many other health problems. The consequences of sowing PFAS into our soils at low levels aren't well studied: they might percolate into groundwater and then people could end up drinking them, and they could also end up in farm products like lettuce or milk.

It's an upsetting prospect that has rattled the industry, even causing some hard-hit plants to stop sending biosolids to farms. But focusing narrowly on biosolids risks missing the big picture, which is that PFAS disperse into the environment through many routes and are already everywhere after years of widespread use, including in our bodies and in the soil. And they get into biosolids because, via the sewer, they've entered a system that wasn't designed to handle them in the first place, and doesn't yet have proven technologies for doing so. That's why many scientists, environmentalists, and industry leaders now argue for cracking down on these and other potential contaminants at their source: the industries that make and discharge them.

Plus rejecting biosolids just resurrects the old problem—what to do with the sewage sludge? In Boston, the fertilizer program not only returns valuable carbon and nutrients to the land, but also helps to pay for the whole wastewater system. (They sell the bulk product for less than half of what it costs to make, but relative to the cost of disposing of sludge in another way, Pawlowski says, they're way ahead.) In New York City, meanwhile, some wastewater treatment plants pay to send sludge by train to landfills. In the spring of 2018, one of these trains, carrying about 10 million pounds of sludge, got stuck in Parrish, Alabama, a town of a thousand people, because another Alabama county blocked the noxious load from its intended rail yard. The train cars sat, stinking, next to youth baseball fields for over two months. "It smells like death," Mayor Heather Hall said at the time.

Soon, new technologies such as hydrothermal processing may offer alternatives that are less fraught than those currently available. But there will probably never be a perfect way to manage sludge— and any solution will likely also meet with some measure of disgust and skepticism by the public. Still, Boston Harbor bears testament to the value of seeking ever-more-sustainable ways to handle the solid stuff. After Boston reformed its sewage system, "the harbor got visibly cleaner," Giblin, the ecologist, says. "They had fewer beach closures. They had less shell rot of the lobsters. They had fewer liver tumors" in flounder. And although Quincy is no longer the Flounder Capital of the World, you can still catch fish there.

Clogged Arteries

What did you just flush?

"At the store, they have one-hundred-percent-recycled toilet paper," Marla says. "The worst job in the whole world must be recycling toilet paper."

—Chuck Palahniuk,
Fight Club (1996)

Potty Breakers

In 2017, sewer workers discovered a revolting monster under Whitechapel, a historic district in London known for everything from Jack the Ripper to mouthwatering Indian curry. It was the city's largest-known fatberg, a clog made up of fats, oils, and grease mixed in with all kinds of trash, all of which had been flushed down toilets and washed down drains. Nicknamed Fatty McFatberg by the public and The Beast by the sewer workers, it measured about 275 yards long, which the *Guardian* reported was the same as the city's Tower Bridge, and weighed 140 tons, the equivalent of two Airbus A318 airplanes or nineteen African elephants. It took eight sewer "flushers" clad in full protective gear nine weeks to break it up and suck it out with a tanker hose. Some parts were so hardened that workers needed pickaxes, shovels, and saws to chip away at it.

Unlike fatbergs before it, Fatty McFatberg was in high demand. A portion went to the Museum of London, which created an exhibit around it. Figuring out how to preserve and display this hazardous waste was a world's first. "Fatbergs are a material that's not well understood, chemically or biologically," curator Vycki Sparkes explained. A live-streamed "fatcam" allowed anyone to watch live as one fragment morphed within its clear sealed display case. It sweated out moisture, changed from dark brown to pale gray to medium beige, and hatched flies. Still, people were drawn to it, doubling the usual attendance at the museum. "I could still smell it," said visitor Cathy Holder as she left, looking a little peaked in a video. "It seemed very musty, and it was almost like I could smell the fatberg as it was when they excavated it, and it was quite disgusting."

From the beginning, toilet systems have been bedeviled by trash. In some cases, people find it easier to flush something than put in the garbage bin (like contact lenses); in others, it seems more sanitary (like wet wipes and tampons); in even others (like drugs and drug paraphernalia such as needles, syringes, and Baggies), it's a way to avoid facing legal or social consequences. Sometimes it's just an accident: one expert I talked to told me that her child managed to flush an entire adult-sized shirt down the toilet.

Most sewage systems find themselves playing an expensive and frustrating game of whack-a-mole, inventing and installing new ways to grind up, filter out, and otherwise neutralize unwanted objects and chemicals so they don't mess up plant equipment, get discharged into a water body, or get embedded in biosolids and spread on land.

There are no perfect solutions if we as a society continue our addiction to stuff. For me, this sinks in on a visit to a treatment plant in the north of the Netherlands, where I stand with a cluster of wastewater professionals who are staring at a fifty-foot line of bubbles running diagonally across the canal that transports the plant's effluent out to sea. It is a pilot version of the Great Bubble Barrier, which its

Amsterdam-based developer calls "an elegant solution to stop the plastic pollution from entering our oceans."

If I could look underwater, I would see a perforated tube on the bottom of the canal, churning upward a never-ending sheet of bubbles. And, the company claims, if I could see any plastic bits in the effluent stream, I would watch them hit the bubble sheet and flow up and out toward the canal's bank, where they're caught and removed. The concept reputedly works pretty well for big pieces of plastic, and even plastic pieces as small as one millimeter, but this pilot is meant to test even smaller bits, known as microplastics, which form as plastic products disintegrate or shed from synthetic fabrics in washing machines. One study, looking at Toronto as a case study, estimated that a city could send hundreds of billions of these fibers into the ocean every year, where they can work their way up the food chain, moving from small fish to big tuna, which can end up back on human plates. I hope the bubble solution works, but it would be even more *elegant*, I think, if that plastic weren't in there in the first place.

Ultra-Soft Power

As I cup the dirty white fluff in my hand, it reminds me of nothing so much as the accumulated dust in a vacuum cleaner, and it's hard to believe that anyone would pay money for it. On a warm late July day in 2019, I have come to a smallish wastewater treatment plant north of Amsterdam to check out a project that I first spotted in viral videos. A company was taking dissolved toilet paper out of the wastewater stream and turning it into this amorphous material, which could then be put to use in many different ways—most famously, as a component in asphalt for a bike path. Could there be something, I wonder, more quintessentially Dutch than this unlikely combination of water engineering and cycling?

Before toilet paper existed, people who wiped did so with a variety of tools. In the Greco-Roman period, people would use stones, ceramic bits (called *pessoi*, meaning "pebbles"), and sea sponges on sticks (though there's some debate among scholars about how popular they really were since no sponges have survived in the archaeological record). Archaeologists even found brown lumps of apparently dried poop on a gorgeous crumpled papyrus containing commentary on Homer's *Iliad* in the ancient Egyptian city of Oxyrhynchus. In colonial American times, people used corncobs and pages from the *Old Farmer's Almanac*.

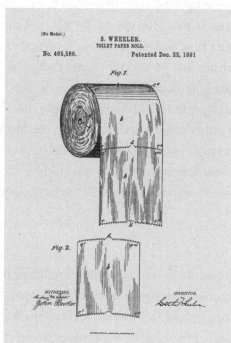

Patent illustration showing a perforated toilet paper roll, 1891.

The Chinese invented paper, so they were probably the first to wipe with it on a large scale, but a uniform, mass-manufactured commercial product first emerged in New York in 1857, when an entrepreneur named Joseph Gayetty began selling boxes of individual sheets of aloe-infused Manila hemp (a fiber made from a relative of the banana plant) that he claimed would prevent hemorrhoids. Tree-based paper on a roll took over when Scott Paper Company popularized it in about

1890. Still, people preferred to buy it discreetly. "Ask for a roll of Hakle and you won't have to say toilet paper!" went one brand's tagline. It was only once indoor plumbing became widespread that people really *needed* toilet paper, since other options at hand were impossible to flush. (And, indeed, in some places even toilet paper is forbidden because the pipes are so narrow. When I lived in Chile, I learned to toss it in the bin next to every toilet. I never quite got used to that pile of streaked and discolored paper, though.) Today, Americans are the world's most voracious consumers of toilet paper, blowing through the equivalent of 141 3.2-ounce rolls of toilet paper per person per year, according to Statista, a provider of market and consumer data. The growing demand for quantity and quality is practically embodied in Charmin's Forever Roll, a single, comically giant ten-dollar roll of toilet paper, launched in 2019, that's equivalent to twenty-four regular rolls. It requires a special stand.

Unfortunately, the softest rolls are made from trees in the boreal forest of Canada, a "vast landscape of coniferous, birch, and aspen trees [that] contains some of the last of the world's remaining intact forests, and is home to over 600 Indigenous communities, as well as boreal caribou, pine marten, and billions of songbirds," according to the Natural Resources Defense Council (NRDC) 2019 report *The Issue with Tissue*. These gorgeous trees produce longer fibers, which make for stronger, softer, more luxurious tissue. While many toilet paper products boast sustainability certifications, trees still get cut down making them, and, as one reporter put it, "destroying forests to wipe our butts can keep you up at night." The NRDC recommends recycled toilet paper—but if you find that your butt can't handle the switch, you're not alone.

The virgin/recycled dilemma—in which virgin inevitably, depressingly, wins—may soon be an issue of the past, thanks to new options. Recently, I ordered recycled rolls from an Australian company called Who Gives a Crap, which the NRDC gives an A+ for sustainability. The company also donates part of the profits to building toilets in parts of the world that don't have them. These rolls weren't scratchy or flimsy like other recycled brands. Even better were the "premium" bamboo rolls, which felt like wiping with silk. The NRDC hasn't rated those but says that fast-growing bamboo can be sustainable, though it's not as good as recycled. A few unbothered souls have turned to the so-called family cloth, which are essentially washable cloth rags that—depending on the type of fabric—reputedly feel even more amazing. And, especially with the addition of bidet functions in new "smart" toilets and bidet seat attachments, there's more and more opportunity to ditch toilet paper altogether. Still, it seems unlikely that these more extraordinary options will catch on anytime soon, since it can be difficult to be out in front of the pack: once, in a futile attempt to get my parents to consider a bidet, I took them to a Kohler store to look at a high-end Numi smart toilet, which also has an automatic lid and heated seat, among other perks. Afterward, my mom laughed that the retractable bidet nozzle was "like a little penis coming out," and you have to believe me when I say that my mom does not usually make comments like that. To her, the toilet was like a weird robotic flasher, one that she would be embarrassed to have in her home.

Colin Beavan, who called himself the No Impact Man during a yearlong experiment in 2007, got so much attention for substituting washing for wiping that he ended up yelling on the air at a BBC journalist who wanted to know details, saying, "I'm sure your listeners are embarrassed for you that you keep pressing me on this issue. Would your mother be proud of you for asking such a personal and embarrassing question within the hearing of millions? Because I'm sure my mother would not be proud of me for answering you." Not a response, I imagine, that created many converts.

There could be another answer. Coos Wessels, a graying engineer with a friendly crooked smile and glasses perched on his head, carefully passes the dirty fluff from his hands to mine. He's the technical director of the company CirTec, from which he has spun off the company Cellvation, in partnership with the company KNN Cellulose, to make the technology that produces this stuff. He explains to me that used toilet paper is a big deal for wastewater treatment plants—nearly three-quarters of the solids that come into a plant are not poop but toilet-paper cellulose. Recovering it there would do more than just reduce wastefulness; it would reduce the load on the plant. Normally, the bacteria just eat up disintegrated toilet paper as they do poop—it's all organic matter, after all. But the more of these cellulose fibers there are in a system, the more oxygen the bacteria need, and the more energy needed to drive the aeration pumps that provide the bacteria with that oxygen. So if plants could take the cellulose out of the system before it gets to the tanks, they could lower their energy bills, as well as increase their capacity overall without building more tanks. And if they could find a way to reuse that used toilet paper, they could save themselves the costs of disposing of it.

When Wessels gives me the tour of the pilot installation, it isn't working because someone has dropped a piece of wood into one of machines. When it's running, it handles the sewage equal to that of about eighteen thousand of the area's residents, which gets pumped out of the main stream entering the plant. The sewage runs through two stages to separate out the cellulose: the first creates a type of turbulence that causes the sand to fall to the bottom and the fats to float on top, and the second is a spinning mesh drum that filters out seeds, plastics, and strands of hair, letting only cellulose fibers through. "Human hairs are the worst there are because we really have thin hairs" that are a lot like cellulose, Wessels says. But there is an important difference between the two: hairs are always bent at the root, so the engineers came up with a clever modification that would catch the hair in the filter but let the cellulose fibers through. It's so clever that Wessels doesn't want to tell me more, lest a competitor steal the idea. At the end, what remains is a product that's about 92 percent pure cellulose and 8 percent undefined material, some of which Wessels guesses are synthetic fibers that come primarily from clothes.

Wessels explains to me that it's actually the high quality of toilet paper that we use—in particular, the length of the fibers in them—that makes this technology possible. "I always thought the fibers in toilet paper would be the lowest grade of material," he says. "It's not." But as they flow through the sewer system, they break apart, getting shorter and shorter. If they were to get too small, the system wouldn't catch them. The product is still "very high quality," he says, better than a lot of cellulose from recycled paper products. But if people decided to suddenly accept rough, less absorbent toilet paper or abandon it altogether, like No Impact Man? Ironically, this "sustainable" technology might not have a business case.

The value of the growing global cellulose fiber market is over $21 billion. It's used in manufacturing insulation, paper, and cardboard and as a source of industrial chemicals such as lactic acid and polylactic acid, a starting material for bioplastics. Cellvation has even figured out that it can break the cellulose down into "sugar that you can use in your coffee," Wessels says. "Nobody wants it, of course."

Concerned about public perception of their product, his company did a little research. At a market, they passed out drinks in paper cups labeled as made with recycled toilet paper from sewage (they weren't in this case, but that wasn't the point). "And most people were, I think, rather positive," Wessels says. "They would drink something from it without a problem." Soon after, however, the company took the cups to an industry symposium. During a refreshment break, they put them out next to plain paper cups with nothing printed on them—and only one person used the recycled cups. "Our own industry didn't take it," he says. "They had all kinds of questions: How is it hygienized? How safe is it?" Famously conservative, the wastewater industry may need extra convincing before it chugs this innovation down, though the public is thirsty for change.

Garbage In, Garbage Out

Growing up in the Philippines, engineer Francis de los Reyes III, now a professor at North Carolina State University, worked for his father's machine shop, at times painting the insides of empty water tanks with highly toxic, anticorrosive epoxy paint. It was dangerous work. "You're twenty feet up in the air, you go down into a small hole," he recalls. "You hold your breath and paint for as long as you can, maybe for a minute or so, and then go up and gasp for air."

Today, de los Reyes develops technologies that could help the people who empty and use pit latrines in low-income regions, but he was always neck-deep in issues of justice and equality. As a student, he was part of "the first what we would call 'people-power revolution'" that drove the dictator Ferdinand Marcos out of his country; even today, the plight of the world's poor is never far from his mind. "My story, part of it is this ever-present awareness of the dissonance of where I came from and where I am now, and I guess you can call it guilt, you can call it whatever, . . . but I see my foot in both worlds."

Starting out as a young academic in North America, he focused on the molecular biotechnology of wastewater treatment for high-income contexts—"the latest and greatest technologies," he says—because that's where the funding was, and, to a great extent, still is. But he had something else in mind. In the late 2000s, he joined the growing number of researchers who were converging around the new field of fecal sludge management, which considers how to deal with human waste in the absence of sewers. They argued that the world's pit latrines—especially the urban ones, which are prone to filling up—could count as safely managed sanitation if professional services would regularly empty them out and transport the contents to a treatment facility.

Over time, it became clear that one of the engineering barriers to this vision is that pit latrines don't just contain pee and poop—they're often also filled with a whole lot of trash, which can clog the pumps that empty them and interfere with treatment processes. That's due to the unfortunate reality that, in many places that have poor sanitation, there's also poor solid waste management. No weekly garbage service, no public trash cans. One small study of eight pit latrines in a settlement in Lusaka, Zambia, found that on average nearly a third of the contents, once dried out, were garbage, including diapers, menstrual hygiene materials, cloth sacks, single-use plastics, rugs, paper, metals, and glass.

Bad sanitation makes this problem worse: widespread drinking-water contamination leads people to buy packaged water, and a lack of clean, accessible, private toilets forces people to make do with plastic bags at night, which they then might throw into the latrine.

In 2011, the Gates Foundation put out a request for proposals for new technologies for emptying pit latrines. De los Reyes and a colleague, who was teaching a senior engineering design class, decided to assign the idea to the students in the class. One student design was interesting enough that they submitted it to the Gates Foundation, which gave them an initial grant to develop the idea, and since then they've just kept making it better and better. After nearly a decade of development, he says, the design now "looks very, very different."

De los Reyes has learned that not all pit latrines are alike. Over the years, he has developed his own "typology" of them. In India, the contents are wet because it's a washing and flushing culture, but there's not much trash because the pits tend to be set off to the side (like septic tanks), so that there's not a big gaping hole to throw things down. "Any pump can [empty] that," he says. On the other side of the spectrum are dry, trashy latrines, like those installed near Durban, South Africa. "They're very efficient with just long shovels and rakes," he says. "And then there are the pits in between; they can be wet, they can be dry, but there's still a lot of trash, or the trash varies."

It's in parts of Africa, in particular, where the pits are both wet and full of trash, that he thinks he can make a big difference. There emptiers with manual or vacuum pumps must go "fishing" before they suck out the latrines. "They literally put water into the pit, they stir it up so that the trash kind of floats, and they use all kinds of manual tools like grabbers and hooks to pull out the trash," de los Reyes says. "You will see mounds of this by the worker's foot; it's really dirty, yucky stuff, and it may take them hours." Fishing is unsafe and also a time suck. In some places, sanitation workers even climb down into pits themselves.

Engineers designing pumps to empty latrines—it's an active area of innovation—can tackle trash in one of two ways. One, they can crush it, but that takes a lot of raw power. "There's so much energy that you need to put into the system to, let's say, destroy a pair of jeans," he says. Two, they can filter out the trash, but the filters suffer from clogs. In the end, after many iterations and prototypes that have been tested in latrines in India, Malawi, South Africa, Zambia, Kenya, Rwanda, and Madagascar, his team came up with a "self-cleaning filter": a perforated pipe protected by a rotating, corkscrew-like auger. They've attached it to a small vacuum unit that can fit down the narrow alleys of urban slums, and they also have a version that works with full-size vacuum trucks. They call it the Flexcrevator.

The Flexcrevator is not only far from the biotechnology work that de los Reyes does for advanced wastewater treatment; it's also far from the futuristic toilets that many Gates grantees are trying to develop for low-income contexts. But he hopes that it will help make good fecal sludge management—and safer conditions for sanitation workers—a reality for people in the near term. "This is a transitional technology, and what I mean by that is there should be a future when there are no pit latrines to be emptied. But that future is still twenty, thirty years away, and, in the meantime, you have 1.8 billion people using pit latrines, and that's not going to go away anytime soon."

Hot Stuff

The hosts of the Olympics often use the spectacle as an opportunity to showcase achievements. For the 2010 winter games, Canada delivered, breaking the record for the most gold medals ever won at a single winter games. And Vancouver did, too, turning the Olympic Village into the first urban site to recover heat from untreated sewage in North America.

People don't flush just water down the drain—or pee and poop, or whatever else—they pour heat down there, too. Think hot showers, washing machines, dishwashers, and pots of boiling cooking water. As a result of all this waste, sewers are warm: wastewater can leave homes very hot, though by the time it mixes with other wastewater and enters wastewater treatment plants, it's usually much cooler, perhaps around sixty degrees Fahrenheit. From a physics perspective, reclaiming this low-grade heat can be difficult—after all, it's too tepid to heat water to the temperatures people need in their homes and businesses, and it's definitely too chilly to turn a turbine to generate electricity. But there is a common, tested technology, called a geothermal heat pump, that can upcycle at least some of that heat. Generally speaking, heat pumps work like refrigerators in reverse, circulating a fluid refrigerant that boils at relatively low temperatures. That fluid can absorb some of the heat of sewage, evaporating into a gas inside the pipe that carries it. Then the gas goes through a compressor that intensifies the heat to a level that people can use—sometimes close to two hundred degrees Fahrenheit—which a heat exchanger transfers to a distribution system. Meanwhile, with the help of an expansion valve, the gas returns to a cool liquid and continues to cycle through the pump. Heat pumps can provide heating without using any fossil fuels, if powered by electricity from renewables, which is why many believe that they are essential tools for combating climate change.

Geothermal heat pumps tend to draw heat from the ground under buildings or from nearby water bodies. (In a fairly typical installation, my grandparents, who lived in a rural area, had one that grabbed heat from a small lake.) But in cities, sewers are even better sources of heat, since they are ubiquitous, warmer than the ground, and relatively close to the surface. Oslo began putting heat pumps in its sewers decades ago to power its district heating systems, which circulate heat and hot water throughout whole neighborhoods. The first, in the area of Skøyen West, was built in 1981 and 1982, in a cavern next to a new sewer tunnel nearly a thousand feet underground. It shut down in 1996 because it had technical problems and wasn't profitable, but the city resurrected it nearly ten years later—with new heat pumps—thanks to funds bookmarked for climate-friendly projects. Today, some 5 million gallons of uncleaned, untreated sewage flow through them each year, meeting the heat and hot water demand for the equivalent of thirteen thousand apartments (additional heat comes from waste-to-energy plants, as well as wood pellet, biogas, and bio-oil combustion). It was, and may still be, the largest heat pump in the world using untreated sewage.

Despite this example, engineers elsewhere haven't been too keen to put the devices into untreated wastewater. Since the gross and chunky stuff in there tends to clog and foul equipment, clever engineering and regular cleaning are required. And there are other potential problems: lowering the temperatures in the sewers could lead to the buildup of fats, oil, and grease on the walls, leading to more fatbergs and affecting the performance of bacteria in downstream plants. Some cities, mostly in Europe and Japan, have placed heat-pump installations on the downstream side of wastewater treatment plants, where the temperatures are lower, but the water is clean. Those plants, however, are often outside of urban areas, which means that the heat can't be put to much use beyond the plant's own buildings and treatment processes.

For the 2010 Olympic Games, Vancouver built the Olympic Village, where the athletes stayed, on a former industrial waterfront site called Southeast False Creek. A showcase of environmentally sustainable concepts, it would continue as a dense, mixed-use neighborhood after the torch got passed to the next host city. At first, they wanted to produce the heat for the centralized heating system from local waste biomass, such as wood waste from parks, but the neighbors worried about air pollution. So they turned to an existing sewage pump station in the area, already slated for an upgrade.

Planners from Vancouver went to tour the few facilities in Europe, including the one in Oslo, and decided to replicate that design at a smaller scale. They couldn't find a European company that wanted to go through the hassle of obtaining Canadian permitting and approvals to sell them the equipment for a one-off project, so they commissioned custom parts from local engineers and manufacturers. To increase the curb appeal, they turned the five stacks from the facility into a public artwork, topping them with LED lights that show how much energy the plant is producing at any moment.

To meet its low-carbon target, the city-owned, self-funded False Creek Neighbourhood Energy Utility (NEU) aimed to provide 70 percent of the space heating and domestic hot water demand with the heat pump. (The remaining 30 percent of the heat could come from high-efficiency natural gas boilers.) They did better than that during the Olympics, as well as after the athletes went home, when the heat pump's capacity was too high for the demand from the empty neighborhood. But, over the years, the neighborhood filled out as planned, and the rising demand meant that the sewage heat portion dropped below the 70 percent goal. Spurred on by Vancouver's declaration of a climate emergency, the utility is now undergoing an expansion of its district heating system, effectively tripling the size from five thousand households to fifteen thousand households, plus a hospital, community center, and other types of buildings. And, thanks to the success of the first installation, they will add more sewage heat pump capacity, which will again put them ahead of their low-carbon targets. "Yes, it's dirty," says Alex Charpentier, a senior energy engineer at NEU, "but it's something that really can be done."

Where district heating systems don't exist, it's possible to put modified, smaller-scale installations in individual homes or buildings. Vancouver has estimated that sewers could heat up to seven hundred high-rise buildings across the region. And in Washington, D.C., the designers of the American Geophysical Union's newly renovated headquarters aimed to achieve "net zero energy," meaning that it would generate at least as much energy as it used through a combination of technologies on the premises. One of D.C.'s combined-sewer lines, built in the 1890s, sits right outside its door, some thirty feet below Florida Avenue. They worked with the German company Huber, carefully drilling into the sewer channel with permission from the utility DC Water, which hopes to sell access to the sewers for this purpose. An additional benefit to heat-pump systems is that, running in reverse on hot days, they can take heat out of buildings and put it back into the relatively cool sewer, providing cooling in the summer. In D.C., where temperatures often reach unbearable heights, this "free cooling" mode means that the AGU building doesn't have to run traditional air-conditioning equipment. So long as the sewer system is below fifty-seven degrees Fahrenheit, which is the case for most of the year, all of the cool that they need can come from there.

Costs will determine whether this technology goes mainstream. Heat from the Vancouver facility is more expensive than if it came entirely from gas, because gas prices are low. But it's cheaper than entirely electric (at least in Vancouver, where electricity comes largely from hydroelectric sources), since it grabs heat for free from the sewers. To the extent that the world follows Vancouver's lead and moves away from fossil fuels, the sewage heat pump will start to look like a good deal.

Junk in the Sewer Trunk

Sewer operators have been discussing fatberg-like deposits since as far back as the 1940s, when it was already clear that grease and trash in the system were a problem without an easy solution. But the situation has gotten much worse. The term *fatberg* itself is new, popularized by London flushers (sewer workers) in 2013. And there's been a great flowering of what's effectively a new scientific field concerning how fatbergs form, how to stop them from forming, and what to do with them when they do—including the possibility of turning them into fuel. After all, FOGs (the industry term for fats, oils, and grease) contain loads of energy.

Scientists used to think that the process of fatberg formation was simple: animal and plant fats cooled in the sewers, solidified, and built up, especially in places where the sewers narrowed, flattened out, took a turn, or had been roughened by corrosion. But in recent years, new research from scientists both growing fatbergs in the lab and analyzing real samples from sewers has complicated the picture. The flowing, churning sewer is like a mad chemist's lab, and at least some

FOG deposits seem to go through a process called saponification, in which they interact with calcium in the wastewater to produce what's essentially a hardened soap, as well as another process called fatty acid crystallization that grows a network of crystals. Researchers are only beginning to understand the role of other compounds in the mix, such as the carbohydrates that come out of rice or pasta water, proteins from meat and milk, and personal-care products such as shampoo. What's evident to any sewer operator, however, is that, once FOGs find a surface to stick to—they like rough ones best—they become like black holes, accreting diapers, tampons, condoms, floss, and any other trash that floats by into their gray, reeking beings. Their consistencies range from squishy, butter-like goop to grainy, sandstone-like rocks.

Though fatbergs sometimes play in the media like a silly horror movie, their consequences are serious. Fatbergs and other FOG buildup can narrow sewer pipes or block them, causing sewage to overflow into streets and back up into buildings. This can be costly, as well as risky for people, animals, and ecosystems that come into contact with the raw sewage. According to one estimate from the United Kingdom, which has old sewer systems, as well as poor FOG regulations, some twelve thousand of nearly twenty-five thousand annual sewer flooding events due to blockages are caused by FOG deposits. Fatbergs have been found to host listeria, campylobacter, and E. coli bacteria and can belch gases, including hydrogen sulfide, which can corrode sewers and asphyxiate improperly protected sewer workers. In China, criminals skim globs of floating fat from sewers to sell on the black market as "gutter oil" to restaurants, with an unknown cost in health and safety to criminals, cooks, and customers alike.

The cleanup costs for fatbergs are high. In the United States, despite better regulations than in the United Kingdom, municipalities spend $500 million to $1 billion per year on clearing fatbergs and other deposits from their systems, according to the International Water Services

Flushability Group. And the FOGs that make it through the sewers to wastewater treatment plants are responsible for about a quarter of sewage treatment costs, according to a European research project.

One way to stop fatbergs is to stop FOGs. These days, people eat more fast food, which is made with lots of oil, some of which ends up going down the drain. Ideally, people would just eat better, with less oil and fatty meat. Beyond that, utilities do their best to encourage home cooks and kitchen staff to wipe off pans and plates with paper towels (something I only started after attending a FOG conference) and to dispose of FOGs with the general garbage. Museum exhibits such as the one in London and another in Detroit help get the word out. Tom Curran, a biosystems engineer at University College Dublin, who goes by Dr. Fatberg, speaks regularly to the media and also performs awkward stand-up comedy about FOGs, including song parodies such as "They Will Block Loos" to the tune of "We Will Rock You."

In many countries, including the United States, restaurants must have—and regularly empty and clean—grease interceptors, which are devices that float the FOGs out of the water entering the sewer system. Regulators check food service establishments and sometimes assess fines, but enforcement can be difficult. In Arizona, the Tempe Grease Cooperative contracts with grease haulers on behalf of all the restaurants that join the cooperative. This saves a lot of money—about 20 percent of the price—because the city buys the service in bulk, and it passes those savings on to the restaurants. And the city can specify what happens to the FOGs—how they're handled, cleaned, and disposed of. Meanwhile, the city of London, Ontario, has focused on homeowner compliance. Barry Orr, the city's sewer outreach and control inspector, and his team came up with a novel outreach idea: a simple cup that residents could put their FOGs in instead of pouring them down the drain. After Orr's team handed out some 125,000 of them, the number of FOG hot spots in the sewer dropped from 101 to 26. Residents can throw out the cups—which is much better than putting the FOGs down the drain—or they can take them to depots, where the contents get turned into energy.

Utilities talk about the three *p*'s that are safe for sewers—pee, poop, and (toilet) paper—but I'm not sure the message has gotten far: in Amsterdam, I went to a wastewater treatment plant serving a million people that had a building the size of a house devoted to screening out large trash that shouldn't be in the sewage in the first place. Speaking for myself, I used to flush lots of trash out of ignorance, including facial tissues, which—*surprise!*—are treated with a chemical binder that makes them break apart more slowly than toilet paper. I flushed tampons because that's what I thought I was supposed to do. Fiddly dental floss always somehow seemed easier to drop into the water than into the trash can, but in a sewer it can wrap around other debris to make a big clump.

I won't cop to flushing a condom, used or otherwise, but you should not do it. I will never forget how, when my biology class went on a tour of my local wastewater treatment plant in high school, we passed a pool blooming with condoms, which had filled up with gas on their way through the system. True to form, I was the kid who asked, "Hey, what's that?," followed by much laughter and blushing by all assembled. Disposable toilet scrubbers, cotton swabs, and cotton balls also belong in the trash, not the toilet. Same for kitty litter, diapers, baby wipes, and paper towels. Sewer operators have nicknames for what some of these can become once underground: cotton feminine products are "muffins," big, shaggy balls of rags are "polar bears" or "sheep," and balls of rags with hypodermic needles sticking out are "porcupines."

But what really seems to feed the fatberg monsters most is a relatively new category of products that are labeled "flushable," which consists primarily of wet wipes advertised as a toilet-paper replacement, and often incorporate plastic or synthetic fibers. The Germans call these *Pumpenkillers* (pump killers) or *Pumpenwürger* (pump stranglers), as if they were serial killers, since they can also destroy the machinery that keeps sewers flowing. While they're flushable in the literal sense that they will go down the pipes, tests that intend to simulate conditions in toilets and sewers show that they don't disintegrate once flushed. Manufacturers, using their own tests, dispute this.

Some wastewater professionals think that better wipe technology or better labeling would help. In the United Kingdom, the brand Natracare was the first to market a wipe that passed stringent tests designed by Water UK, which represents water and sewer companies in Britain, giving it the right to use a new "fine to flush" symbol. Costing about ten thousand dollars a go, the tests check not only for disposability but also for plastics and residuals that could contaminate waterways if they ended up in effluent. But other experts argue that the concept of flushables should be banned altogether, as it could create confusion among users about what's allowed in the toilet. Wipes overall are on the rise: in 2018, the product development company Smithers found that the global wipes market had soared to $16.6 billion. In the meantime, ratepayers are picking up the cost of clearing clogs and investing in expensive new pumps, grinders, and screens intended to deal with this menace, as well as experimental bacterial treatments that eat clogs right in the sewers, and sensors and robots that monitor pipes to see where blockages might be forming.

London's Fatty McFatberg proved to be a leader in more than just size. In addition to media and museums, there was another entity interested in it: the Scottish company Argent Energy. Since the early 2000s, Argent's mission has been to find and process waste fats and oils into biodiesel in order to make the transportation sector less dependent on fossil fuels. Argent started with animal fats but was intrigued when the Scottish city of St. Andrews approached them with a fatberg problem around 2008. The birthplace of golf, the city was experiencing major sewer blockages during the competition season. With all the tourists eating in the restaurants, says Mike Hogg, who's in charge of finding new sources of fatty and oily stuff for Argent, "all the sewers started bubbling up, and there was stink everywhere."

As a trial, Argent agreed to process some of the fatberg at their plant. They could do it, but not easily, as fatbergs contain highly degraded and contaminated FOGs compared to, say, used cooking oil. They had to put off a larger fatberg program, Hogg says, "but it sparked the idea."

Almost a decade later, when sewer workers discovered Fatty McFatberg, the timing was much better: Argent had just opened a new plant in Ellesmere Port, which is about eleven miles south of Liverpool, England. This plant was special because it incorporated a new pretreatment step that could upgrade the sewer FOGs, removing any water, trash, food particles, and chemicals that got mixed in. After that pretreatment, the material could then go through the standard chemical process called transesterification that is used to make most biodiesel, in which fatty acids react with an alcohol. "The pretreatment plant at Ellesmere Port gives us capabilities that very few other biodiesel companies would have in the world," Hogg says. Argent partnered with the city to handle the massive fatberg, and the plant transformed it into more than twenty-five hundred gallons of biodiesel, which then helped power a London bus fleet.

It's a feel-good ending for Fatty, but the fact that Argent can make biofuel from fatbergs should be cold comfort, Hogg warns. While the company continues to process sewer FOGs, it's a far more intensive and expensive way to retrieve the energy in them than collecting them before they go down the drain. "Hunting fatbergs, if you like, isn't a way to feed a biodiesel plant; it's the consequence of sewer abuse," Hogg says. A fatberg, as some have pointed out, is like a heart attack for the sewer. And as with a heart attack, prevention is far easier, cheaper, and more effective than the cure.

Giving a Crap

Come one, come all.

Where, after all, do universal human rights begin? In small places, close to home—so close and so small that they cannot be seen on any map of the world. . . . Unless these rights have meaning there, they have little meaning anywhere. Without concerted citizen action to uphold them close to home, we shall look in vain for progress in the larger world.

—Eleanor Roosevelt (1958)

A Seat of Privilege

The Palatine Hill is one of the most visited tourist sites in Rome, the glorious center of a fallen empire. There stood the luxurious palaces of the elite, and there the elite sat on luxurious latrines—some bathrooms were practically throne rooms. But, as in every era, not everybody was so lucky.

It's 2014, and, in one of my greatest triumphs as a science journalist, I have convinced a dynamic duo of classical archaeologists, Ann Olga Koloski-Ostrow of Brandeis University (known as the Queen of Latrines) and independent Dutch archaeologist Gemma Jansen, to allow me to join them for a descent into the bowels of this famous site. They have secured rare access to a two-thousand-year-old multi-seat latrine, which is now deep underground, at the bottom of a winding set of steps and passages. With just a short time to document it, they complete Jansen's checklist, counting holes, measuring heights and widths, and checking for windows, paint, and graffiti.

If one big picture emerges, it is that this was not a nice bathroom. Yes, in many ways, it is typical of Roman latrines: a large rectangular room with unpartitioned toilet benches lining the walls. The dinner-plate-sized holes in the benches are an odd shape often described as keyhole-like—round on top, with a linear bit that extends toward and down the front of the bench, probably so the users could have access to wipe their undersides with a sponge on a stick or some other implement. But this latrine is unusually large, with about fifty holes. Little attention has been paid to comfort or decorative elegance, based on remaining traces of a plain red-and-white color scheme. And the waits at times were long, which the archaeologists infer from the graffiti that people left outside the door. This latrine was likely the crowded, smelly lavatory for the palace's servants or slaves.

In one sense, toilets are the great equalizer. Everybody poops, of course. But they're also a metric for inequality in any society and in any era. Show me the toilets you use and I can tell you a lot about your status. As Jewish feminist theologian Judith Plaskow writes: "Because the need for toilets is universal, their availability and distribution provide a particularly revealing map of power relations in American society, an everyday lesson in who merits social recognition and whose time is considered to be of value." Put another way, toilets are a key example of the feminist revolution's mantra that "the personal is political." And since the structural inequalities of our societies manifest themselves in our toilet systems, fixing those systems will require not only new technologies but also social change.

From a global perspective, American homeowners are the new

Emperor Neros. While billions of people worldwide still don't have a single adequate toilet, many homes in the United States have undergone what one reporter has called an "opulent bathroomification." Between 1973 and 2019, according to U.S. Census Bureau statistics, the percentage of newly built single-family homes with two and a half or more bathrooms (meaning three or more toilets) rose from 19 to 62. The percentage of homes with two or fewer bathrooms (meaning one or two toilets) plummeted from 81 to 37. In that same stretch of time, the average household size has shrunk from about three to about two and a half members. Altogether, those numbers suggest that Americans prefer at least one toilet for every member of the household—a rather remarkable use of space, since it would be rare that everybody in a household uses the toilet at the same time. (I personally can attest to the American-ness of this trend, since my husband and I have gone to great lengths to find apartments with two toilets in the European cities in which we have lived, where that feature is rare and substantially hikes up rental prices.) The luxury of American bathrooms has also increased. Zillow, an American real-estate database company, found that a bathroom renovation was the second most common improvement made before a home sale, after painting. As Zillow told *The Atlantic*, "a simple bathroom remodel—such as replacing the toilet, adding a double sink, or tiling the floor—carries the best bang-for-buck of any home renovation. At $1.71 in additional home value for every $1 spent, it's three times as cost-effective as a kitchen renovation."

The people at the bottom, meanwhile, know that toilets can serve as a tool of control, punishment, or exclusion—one with potentially life-threatening consequences. Nothing says "you don't belong here" like refusing to put toilets in a homeless encampment or "I own you" like making a prisoner eliminate in a dirty toilet without a door. In the United States, according to a 2016 report from Oxfam America, some poultry workers wear diapers because their supervisors deny them bathroom breaks. And a recent investigation by *Vice* revealed that finding a toilet can be particularly hard for gig workers driving for outfits like Uber Eats and DoorDash: "Restaurants will post signs or tell drivers that bathrooms are reserved for customers only, forcing them to use the bathroom outside or pee in cups in their cars"—makeshift solutions, by the way, that many women would find near impossible. An infuriating new product, called the StandardToilet, is angled slightly forward to discourage workers from sitting on it for too long, making the restroom, according to one expert, just "another place where people impose the very capitalist idea that people should always be working."

This 1975 image was part of a slideshow used to familiarize public health inspectors with the unsanitary conditions commonly found inside migrant worker camps.

As for public toilets, they too often fail people, especially the vulnerable and marginalized, either because they don't exist or because they don't accommodate all that people might need or want to do in them. Those activities include some of life's most stigmatized: changing menstrual products, vomiting, crying. Not to mention helping others who need help in the toilet because of age or disability, being helped, adjusting clothing, changing diapers, breastfeeding, administering injections, spending time alone, sending a text message, checking

Facebook, and escaping a threat. Restrooms are even places where people seek out and have sexual encounters, sometimes because it's a thrill but also sometimes because it's the only private, safe place they can find. Worldwide, it's almost inconceivable that a third of all children lack a basic sanitation service at their school and a fifth of health-care facilities have no sanitation service.

The labor and profits that come along with our toilet systems are not fairly distributed, either. On the one hand, a study from Australia found that women are more likely to scrub toilets while "men are more likely to do the episodic chores such as mowing the lawn or changing the light bulbs." On the other hand, men are more likely to have paid jobs in sanitation: a 2019 report from the World Bank, which focused on low- and middle-income countries but also echoes the reality in many high-income countries, found that just one in five workers in water utilities worldwide are women, with one in three utilities having no female engineers at all.

The Italian artist Maurizio Cattelan best captured the toilet as the ultimate symbol of inequality with his 2016 work *America*. It is a fully functioning toilet made of eighteen-karat gold, designed to look like a standard institutional Kohler toilet with a simple handle flush. It was first plumbed into a fifth-floor restroom of New York City's Solomon R. Guggenheim Museum, where more than a hundred thousand people waited some two hours in line to use it. It is "one-percent art for the ninety-nine percent," Cattelan joked.

As it happens, *America* is far from the only gold toilet in the world. One very rich man, Hong Kong jeweler Lam Sai-wing, commissioned a toilet made of solid gold, inspired, in a head-scratching kind of way, by Lenin's claim that building public toilets from the metal would be a fitting way to memorialize the lives wasted in war "for the sake of gold." Gold-colored or gold-plated toilets have become a gaudy symbol of affluence readily available on Alibaba.com and other international shopping sites for people who are rich or wish to look and feel so. And at least one famous American shares these tastes, as the Guggenheim observed: "The aesthetics of this 'throne' recall nothing so much as the gilded excess of [Donald] Trump's real-estate ventures and private residences" (an excess that reportedly does extend to the bathrooms). In September 2017, the museum offered to loan *America* to the White House for Trump's private rooms, instead of a requested Vincent van Gogh painting.

In 2019, *America* crossed the Atlantic to the United Kingdom for an exhibition of Cattelan's works in Blenheim Palace, where it was installed in a wood-paneled water closet once used by Winston Churchill. After just a short time on display, all 227 pounds of it were stolen, presumably for the raw value of the gold. The remarkable theft involved yanking the toilet from the building's plumbing, which caused structural damage and flooding. Cattelan seemed less than concerned: "I always liked heist movies and finally I'm in one of them," he responded.

Equality. Period.

I still remember the shame and humiliation. As a twelve-year-old, I left home for an evening class. Monthly bleeding was still new to me, and I hadn't mastered the hypervigilance that's required during those days—always planning ahead, always changing tampons given the opportunity, always packing a backup pad. (I still haven't.) Sitting in the classroom that day, I felt the familiar wetness on my inner thigh, and I'm sure my face went white, thinking of the dark stain that could be on the back of my jeans. The stain, as it happened, was enormous, and I ran to the administrative office, sobbing, and called for my mother to pick me up.

Few topics are more taboo than excretion, but menstruation may be one of them. I tried to forget that incident as quickly as I could, and I can still feel the heat in my cheeks as I write it down now. But, in recent years, a growing group of researchers and activists have decided to shine a frank spotlight on menstruation, realizing that understanding and supporting adolescent girls and others who menstruate during their periods can have an outsize impact on their well-being. After all, women menstruate for an average of 2,535 days in a lifetime, which is almost seven years. Menstrual activism is "a movement growing at a break-neck speed," writes women and gender scholar Chris Bobel in her book *The Managed Body*.

A major rallying cry for the movement has been what's come to be called period poverty. While managing menstruation is hard enough, it's impossible when you don't have ample, good-quality menstrual hygiene materials. And since tampons, pads, and other products are expensive, costing up to $120 per year in the United States, which amounts to more than $4,000 over a lifetime, many menstruators struggle to afford them. The consequences of this can be dire. They might use unhygienic materials like socks, newspapers, and tissues, or rags and leaves. They might stay home from school or work. Or they might engage in risky behaviors, such as shoplifting, to acquire the products. In rural western Kenya, researchers found in a survey of 3,418 menstruating females that a tenth of their fifteen-year-old female subjects reported engaging in transactional sex to get money for menstrual pads, which would increase the chances of sexually transmitted infections and pregnancies. (For participants older than fifteen, the prevalence of reported transactional sex for this purpose was substantially lower, and the authors urge caution in interpreting the data.)

In some countries, activists have focused on what they've come to call the tampon tax. The term is a bit confusing because the tax on period products is the same sales tax that is on many other products. But most tax codes exempt items that are deemed necessary. In various American states, those items range from prescription drugs to dandruff shampoo and lip balm, not to mention oddball products like marshmallows and cooking wine and snowmobiles. (Sounds like a nice tax-free party, doesn't it?) Period products have not been on those exempted lists, however, which amounts to a sexist policy, since the tax is levied nearly exclusively on women, as well as nonbinary and trans people who menstruate. The weight of the tax burden also falls most heavily on poor people, since, unlike income tax, sales tax gets calculated the same, regardless of whether the purchaser is rich or poor. (If you've heard of the so-called pink tax, that is a separate issue, since *the pink tax* is not a term for an actual tax but refers to the higher prices that companies slap onto products for women, from razors to deodorant.)

Kenya lifted its tampon tax in 2004, and Canada, India, Malaysia, and Australia have done so more recently. In the United Kingdom, university student Laura Coryton started a campaign called Stop Taxing Periods in 2014. An online petition through the platform Change.org garnered more than 320,000 signatures, a success fortified by protests and social-media outreach. It didn't take long for politicians to get on board; the problem, however, was that the European Union's rules wouldn't allow member states to reduce any sales taxes below a required minimum of 5 percent. Coryton's campaign got caught up in the Brexit debate, with Leave politicians promising to drop the tampon tax if Brexit succeeded and Remain politicians promising to work with the European Union to change the policy. (When Prime Minister David Cameron said "tampon" in Parliament, it was the highlight of Coryton's life, she has said.) In the end, both avenues have borne fruit: post-Brexit Britain decided that the tax would end in 2021, and the European Union has also agreed to change its tax policy, allowing tax-free items by 2022. Some other European countries, many of which tax period products at a whopping 20 percent rate, are preparing to scrap their tampon taxes soon after that.

In the United States, sales tax rates depend on states and municipalities. In Utah, taxes on tampons and other products amount to about $1 million, while in larger California, it's about $20 million. Lawyer Jennifer Weiss-Wolf, who started the PeriodEquity.org campaign in 2015, says that focusing on equality instead of poverty has allowed the issue to cross party lines. In 2016, President Obama weighed in during an interview, thrilling advocates and engaging new supporters of the cause when he said, "I have no idea why states would tax these as luxury items. I suspect it's because men were making the laws when those taxes were passed."

Some states have passed laws eliminating the tax. Still, there's a long way to go: in early 2020, a Tennessee lawmaker pushed back on the effort to include period products in a three-day tax holiday, expressing concerns that "there's really no limit on the number of items anybody can purchase," he said, evidently oblivious that a menstruator doesn't need—or want—an endless supply of tampons.

For some, however, a tax break isn't enough to ensure they get period products. In 2020, Scotland became the first country to provide free products to "anyone who needs them"; England and Wales now require that schools offer them for free to students, as do a few U.S. states. The governments of Kenya, Uganda, Zambia, and Botswana have taken steps in that direction. Worldwide, programs have made other attempts to distribute period materials for free, especially to girls in school, in the hopes that they won't miss days or drop out. India's famous Pad Man, Arunachalam Muruganantham, was a school dropout who, bothered by the old cloth his wife used, invented a tabletop pad-making machine, which he then practically gave away to small nonprofits and women's self-help groups so that they could create small businesses around them. At first he endured ridicule, but in 2018, Bollywood released an adulatory biopic based on his story.

Then there is the question of what else menstruators need: a decent place to change and dispose of or clean their products—and this is where toilets come into this story. As a Peace Corps volunteer in an Eritrean school in the late nineties, Marni Sommer noticed that her lone Eritrean female colleague missed a few days of work the first month, then again the next. She asked the male teachers about it, but they just said that the teacher was sick. After three months of this pattern, Sommer put it together: "Finally, I was like, *oh*! And then they're like, *yes*." The only toilet in the school was a less-than-private affair in the teachers' room. "There was a little corner and that was the bathroom and none of the Eritreans would use it," she recalls. "I never talked to her about it." The students had it worse, since there was no toilet for them at all and they had to relieve themselves in the nearby forest. "I started wondering at the time. . . . Are they dropping out because of [their periods], or is it that they're getting married because they get their period? And so I just came back from the Peace Corps with all these questions."

Now, as a public health expert at Columbia University Mailman School of Public Health, she studies how to make toilets work for menstruating women. Female-friendly toilets, as Sommer and her colleagues now call them, should have some basic features: doors that lock from the inside, water to wash blood off hands and clothing, a hook or shelf so that women have a place to put clean menstrual products other than on the muddy floor, and a place to dispose of used products.

Programs that provide pads and toilets don't always work as planned, however—often because people who design the programs don't think beyond the product itself to the needs and desires of the people they aim to help, or to their social context. In Bangladesh, a program failed to make sure that girls had a place to throw out their pads once they were done, so the pads ended up strewn outside. In Tanzania, a program failed to reach out to parents, so Maasai girls didn't take their pads home for fear of their parents' reactions to the taboo items. In one pilot by the International Rescue Committee in a refugee camp in western Tanzania, a team created a special menstrual hygiene management toilet facility, but women didn't want to use it because they didn't want others to know that they were menstruating. In Cambodia, a program failed to overcome girls' hesitancy toward menstrual cups donated under a "buy one, donate one" scheme, so the items ended up for sale to foreigners at a hip downtown cafe. In rural Uganda, a program taught women to make their own pads, but the women told organizers that they would rather learn to make oil and soap, so that they could buy the menstrual products that they preferred.

"At the root of most of the challenges associated with menstruation is menstrual stigma," Inga Winkler, a legal scholar at Columbia University's Institute for the Study of Human Rights, has explained. "We can't address menstrual stigma by simply providing pads. We can only address menstrual stigma by speaking about menstruation much more openly, by showing what menstruation is about, and by changing societal norms." In many cultures, those norms mean girls and women must limit their movement during their periods—say, by staying at home or avoiding religious spaces. In Nepal, some girls and women are banned from their homes during their periods and sleep instead in unsafe huts (although this practice has been outlawed, it still continues in some places). But even in cultures in which that's not the case, menstruators often avoid activities, like sports and swimming, for fear of leaks, and don't feel comfortable seeking help when they have symptoms such as heavy bleeding and pain.

In addition to providing products and lifting taxes around the world, Winkler believes that activists should also aim to normalize periods by presenting the diversity of menstrual experience—positive or negative—instead of the simplified version often seen in the media and discussed in schools. Artists have weighed in here, starting with Judy Chicago, who in 1971 created a photolithograph of a woman removing a tampon. India-based artists are now in the mix: Lyla FreeChild paints with menstrual blood, and Sarah Naqvi embroiders underpants, pads, and tampons with red thread and beads. Said Naqvi to a reporter: "It is tough enough to deal with menstrual cramps, and if you're asked to hate yourself it becomes more challenging."

Passing the Smell Test

In 2012, the Gates Foundation, in its quest to get safe toilets to the world's poor, approached a fragrance company with a challenge:

Could the company's scientists help make toilets designed for poor settings smell better—*much* better? No matter how safe they may be, toilets that collect waste on site can emit unpleasant odors. Smell is an ancient, primal, subjective sense, which we often process subconsciously, so science doesn't always have a clear view of how it influences behaviors. Still, one effect seems obvious enough if you think about it: If a toilet stinks, people might not use it. If it smells great, it might even beckon.

This meant developing a whole new chemistry of stench, or malodor, control. Firmenich is the world's largest privately owned fragrance and flavor company, founded in 1895 in Geneva, Switzerland. If you haven't heard of it, that's because it doesn't market scents and flavor to you; it sells them directly to the companies that make the shampoos, lotions, cleaning supplies, and other products that you buy. "We're kind of a company behind a company," Sarah d'Arbeloff, director of global technology business development, said in a talk.

The goal was a fragrance that would first block bad odors and then introduce good odors on top of them, rather than a strong, appealing fragrance that would overpower a bad one. Christian Starkenmann, who has since retired, was one of the Firmenich chemists leading the project. He was reassigned to it because, unlike most fragrance scientists, who try to generate lovely smells, he had been working to understand smells such as onion breath and body odor. This work left him with a remarkable ability to describe the components of bad smells the way oenophiles describe the rich flavors of a particular red wine. Even so, he had to learn to love working on "shit and urine."

First, the team had to understand the stink that they were fighting. Science had long before discovered the main component of fecal odor, thanks to the pioneering work of Dr. Ludwig Brieger in Switzerland, who, in the 1870s, distilled more than a hundred pounds of fresh feces from patients—the "good stuff" from healthy people who had injuries and not those who had an illness—in his lab. From it, he extracted a compound that he called skatole, from the Greek root *skat-*, for excrement. He determined that it was a product of digestion, observing that diarrhea lacks it because the transit time of the food through the gut is too short for the digestive system to produce it. Skatole is particularly important because it has a low "odor detection threshold," which means that people can smell it in very low quantities, unlike some other compounds, which require a large volume in the air before people notice. It's also, as it turns out, potentially toxic and cancer causing, but thankfully we don't produce enough to poison ourselves. Weirdly, it doesn't smell like poop but instead like "moth balls or a slightly wet basement," according to Starkenmann, and is a component of the fragrance of orange blossom and jasmine. It's only when skatole combines with other compounds that the characteristic stench of feces arises.

The stink of latrines is something even different, and more complex, from that of fresh feces and depends on factors such as the type of pit, how often it's used and emptied, the room around the toilet, and even cleaning supplies. To study the "headspace" of these heads, Starkenmann and his colleagues went on what he calls "a world tour of toilet smell"—a tour that I envy but few others would want to take, and which was eye-opening for a French-speaking Swiss scientist accustomed to orderly, clean laboratories.

The first stop was Durban, South Africa, where the team sniffed some of the city's well-known ventilated improved pit latrines, often referred to as VIPs, which collect feces and urine in separate places, creating a kind of humanure out of the former. When VIPs were functioning well, the team found, they had a "faint, barnyard odor," but they could also produce awful smells, such as that of stale urine, as well as a "stinky sewer smell" of "sewage, bad eggs, cabbage, and sulfur" when the pits got full of water. In Kampala, Uganda, the team saw two more toilet styles: one was an underground plastic cylinder connected to the bowl by a narrow pipe, to keep out trash; another was an elevated toilet, positioned over a very stinky ground-level tank and reached by a rickety ladder.

Their third stop was Nairobi, Kenya, where the communal Fresh Life toilets of the company Sanergy (which we already encountered in the context of black soldier flies) impressed them for their efficacy and sustainability, but not especially for their smell. Underneath each toilet was a concrete space containing one drum for feces and another for urine. Although the toilets were cleaned regularly and emptied daily, poor ventilation in some of the models meant that odors lingered.

Their last stop, in Pune, India, presented the worst of all smells, coming from large communal toilet blocks. A men's urinal room reeked so badly of ammonia that Starkenmann could hardly breathe. Squat toilet rooms blasted him with the stink of shit because of residues that coated the bowls. There were no septic tanks or sewers—the waste flowed into the river, which smelled worse than the toilets, "not only from all the toilet waste but also from other garbage, including dead animals, which found its way into the water system."

In each of these locations—sixteen latrines in total—they did their best to sample the odors. Smell scientists often use a fiber coated with a special polymer that can absorb compounds from the air. But they thought that some of the key compounds would have very low odor-detection thresholds and might be present in such low quantities that the fiber wouldn't pick them up. So they settled on a solution in which they took samples of the contents of the toilets (the pits where possible, or otherwise the residue on the toilet bowls), put those samples in small vials, added helpful chemicals, and then put a fiber into the vial to extract the compounds. With high-tech lab equipment, they then analyzed the fiber. Later, they realized that this strategy had still missed some important constituents of the odor, so they went back and used a different technique, pumping latrine air for three hours through a liquid baking-soda solution, which they then percolated through some cartridges that could capture the molecules.

In the end, the team found about two hundred chemical constituents, which they narrowed down to five key molecules that could re-create the stench of latrines, Goldilocks-style, just right: "not too like fresh poo, not too cheesy, not too barnyard-like, and not too sulfury." What they then needed was the opposite of that poo perfume—a sort of antimatter to neutralize the bad odors, known as a "white fragrance." Testing the latrine molecules against a wide array of mouse olfactory cells and then matching the mouse cells to their human counterparts in a database, the team found the receptors in the human nose that respond to the stinky molecules. Then they found odorless molecules, called antagonists, that attach to the same receptors, blocking the nose from smelling them.

Since the antimatter fragrance wasn't perfect, the team then planned to blend those antagonists with good smells that would cover up any remaining odor. But which good smells? Because smell is subjective, different cultures, and even people within those cultures, can perceive odors differently. On arrival in India, Starkenmann remembers, he sniffed the air and remarked on the unpleasant tang of feces. For his seatmate, a native of India who had been living abroad, it smelled comforting, like home. Still, Starkenmann's team found in their travels that most people rated fecal smell as offensive and preferred other fragrances for their toilets. At first, the team thought that an obvious choice would be jasmine, since the molecule skatole is a component of jasmine smell, so you could just layer the other jasmine molecules on top. But in some countries, they found, jasmine is a smell for religious ceremonies, not toilets—and just imagine your toilet smelling like a church! In still other countries, people want toilets to smell like tar, since they're used to cheap cleaning products made from charcoal, which smell strongly of tar and kerosene. The scientists added their formulations to cleaning products or impregnated them into scented pads, which last for a few days, depending on the conditions in the toilet.

Firmenich is part of the Toilet Board Coalition, launched in 2014 by five big corporations and an Indian charitable foundation in order to bring business solutions to the global sanitation problem. You've more likely heard of some of the other founders, such as Unilever, a European personal-care and cleaning-product manufacturer, and

Kimberly-Clark, an American paper goods manufacturer. Instead of being a great big black hole of costs, the coalition argues, sanitation—even when delivered to the "base of the pyramid," the poorest of the poor—can generate value. One way is through capturing "toilet resources," which are the nutrients, energy, water, and other valuable materials in sewage and fecal sludge—resources, the coalition adds, that will only increase with population. There is also potential profit in providing toilets and the products related to them, as well as in the collection and use of sanitation-related data for a wide variety of purposes. In one study, the coalition estimated that the total value of the sanitation economy in India will be $62 billion by 2021.

In the fragrance industry, clients typically send companies like Firmenich a brief about a fragrance they need for a product. For the new malodor technology, however, Firmenich had potential clients pitch them on how they would use it to make an affordable product that would target the base of the pyramid. Before the clients could get a sample, they had to sign an agreement not to analyze the scent or try to seek a patent for it, since a patent for any one client would prevent others from using it and therefore limit its reach. "With some of our global clients, it is the norm that they would automatically try to figure out what's in there," d'Arbeloff said, "so this took a lot of negotiation." Firmenich also wanted to make sure the fragrance would only be used in conjunction with actual cleaning products so that it wouldn't mislead people into thinking an unhygienic toilet was clean.

At the right price, the product could help public toilet businesses. With the consulting and market research firm Archipel&Co., Firmenich ran a trial in public toilets in India, finding that they could increase traffic up to 16 percent over a six-month period with regular cleaning, odor treatment, and community engagement campaigns—a combination that they admit is intensive. In Zambia, entrepreneur Mwila Lwando, who provides paid public toilets branded Live Clean at markets in Zambia, got samples of the scented pads. Lwando is not one to pull punches if he has something critical to say, but he gushes about the pads. "The scent is so strong," he told me. "It has made people always ask, what's this?" People even started stealing them—women to put in their handbags, bus drivers to put in their buses. When the pads have finished their duty in the toilets, Live Clean gives them away to customers and workers as a kind of bonus.

A malodor perfume will not solve the world's sanitation problems, Starkenmann says. "It's a detail." But it's a detail that, in his travels, he has found makes a difference to many of the people he met, some of whom had to use facilities in abysmal conditions that still make him angry to think about. A nice-smelling toilet makes people feel cared for and respected, he believes. And that feeling is something that, like safe sanitation, he thinks everyone deserves.

A Public Domain

On a spring day in 2018, two young Black men entered a Starbucks in Philadelphia, where they planned to have a meeting with someone who hadn't yet arrived. They asked to use the restroom, but, since they hadn't bought anything, the staff refused to give them the key. Then they sat down and waited. Within minutes, a manager called the police, who arrested them. A video of the arrests went viral, and protests erupted online and in the store. The incident rightly became part of a larger narrative about racism and policing in America.

It's easy to see Starbucks as the enemy in this story—and following the outcry, the company took several measures: it settled with the men for an undisclosed sum, closed all stores for "anti-bias training," and changed its policy to allow anyone to use the toilet, including those who have not paid. But the problem here is not just Starbucks, or even endemic racism. There's another underlying systemic issue— namely, that it's hard to find a public toilet these days, so businesses are filling in the gaps, poorly. In Washington, D.C., an advocacy group performed an inventory and found only two round-the-clock public restrooms in the whole city. They also sent representatives to eighty-five downtown eateries over several years, asking to use the bathroom.

In 2014 and 2015, forty-three of those said yes; by 2017, only eleven did. "We don't want to become a public bathroom," admitted Howard Schultz, then the CEO of Starbucks, in his public mea culpa. "But we're going to make the right decision a hundred percent of the time and give people the key."

In the middle of the twentieth century, public toilets were relatively common in the United States. For the most part, people had to pay for the stalls, which opened with coins, while urinals were free. But with rising feminist sentiments in the 1970s, people started to recognize this arrangement as sexist, since it meant that women had to pay, while men did not. Over the following decade, the grassroots organization Committee to End Pay Toilets in America succeeded in getting bans on pay toilets in many states, and the late March Fong Eu, California secretary of state from 1975 to 1994, smashed a porcelain toilet with a sledgehammer on the steps of the State Capitol.

The banning of pay toilets had an unintended outcome. Instead of free public toilets for all, public toilets pretty much disappeared from American cities, since cities found it expensive and difficult to keep them safe and clean. (In Europe, where there was no such movement, pay public toilets persist, as they have since as far back as Roman times.) And with any remaining bathrooms unsafe, unclean, and in disrepair, they got a reputation for attracting illegal activity. Ironically, this situation, like that of the pay toilets that feminists protested, falls more heavily on women, who need to use the restroom more often than men, especially if they are pregnant or menstruating.

In response to growing grassroots concerns about the lack of public toilets, some have worked to improve their design. High-tech automatic toilets, which clean themselves, hold promise but have often run into trouble, breaking down and sometimes even trapping people inside. On the other hand, the low-tech Portland Loo, from a crack team working for the Oregon city, has earned a fervent following. This "soulless receptacle for bodily waste," as one admiring reporter put it, is small enough to fit on the sidewalk or in a parking spot. The "defensive" design is maximally inhospitable to people who would want to commandeer it for illicit purposes: no running water inside (the tap is outside), no mirror, metal bars instead of walls at the top and bottom to reduce privacy, a graffiti-resistant coating, and walls of indestructible stainless steel. Some experts advocate bringing back the public pay toilet, for which cities should charge men and women equally. For the poor, cities could implement free token schemes—"toilet stamps," if you will.

But what if businesses like Starbucks, instead of covering ad hoc for a city's restroom deficiencies, could become part of a larger solution, providing convenient public toilets that are also comfortable and free? In Bath, England, a large old public toilet was big enough that the city was considering turning it into a one-bedroom house. Then two local women proposed a better idea: turn it into a gift and housewares shop with an attached toilet, which they would maintain and open to the public during business hours as part of their lease. The city agreed.

Similarly, there's the City of London Community Toilet Scheme and Nette Toilette in Germany, along with like-minded projects that have been popping up worldwide. The idea is simple: a business like a cafe or parking garage agrees to open its toilet up as a public convenience, advertising it with a sign in the window and on an app, and in return the city pays a small stipend to help with upkeep. Some businesses benefit from the free advertising and increased foot traffic that the scheme brings them. As part of an effort to fund new restrooms, Washington, D.C., is planning to run a pilot like this, with an estimated incentive of about two thousand dollars per year to businesses. On the downside, however, one wonders whether businesses can be trusted to treat everyone equally; these types of programs would require oversight if the neediest and most vulnerable won't be excluded anyway.

In Pune, India, toilets downtown are scarce, especially for women. And, because of the density, there is no room for new buildings. Rajeev Kher and Ulka Sadalkar already had a business providing portable sanitation to worksites and festivals—a profitable enterprise, but not aimed at the poorest. Working with the Pune Municipal Corporation, they came up with the idea of converting thirteen-year-old city buses, which were waiting to be scrapped, into sanitation "health centers" for women. They call them Ti, which translates to "she" or "her" in Marathi, a local language, and also stands for "toilet integration." A local company paid for the refurbishment, and as of 2020 there were a dozen operational buses in the city of Pune, with more planned in other cities. Trained attendants keep the cheerful black-and-white interiors clean—an improvement over the city's other public toilets that regularly go out of service or are plain disgusting. They offer two kinds of toilets—"Western" and "Indian"— along with washbasins and diaper stations.

So far, it sounds like a typical public toilet—a costly affair. But they've found other ways to boost their revenue by adding pay services such as Wi-Fi, charging stations, laundromats, and cafes. Sensors in the bus (developed by MIT) alert attendants to when the temperature and humidity change, which warns them when the toilets might get stinky, and scent pads from Firmenich keep them smelling fresh. There's advertising space on or inside the buses for sale. Kher has said that the inspiration for this comes from the coffee giant itself:

> We've perceived it as a Starbucks or a McDonald's. Many people go there when they have to use the toilet because it is clean, and then end up having a burger or a coffee. Now, move this concept to the end of the pyramid and apply this same logic. Everyone thinks the same way, only the price points are different. You can call our [Ti] bus a ladies' room or a lounge where women can freshen up and pick up a kokum sherbet on [their] way out.

While Starbucks may not want to become a public restroom, the public restroom wants to become Starbucks.

Nuke the Powder Room

On the cusp of graduation, Seb Choe, an architecture major at Columbia University in New York, attended a 2017 lecture on public restrooms by architect and Yale professor Joel Sanders. For Choe, a trans nonbinary person, the concept of male and female restrooms had never worked. "It's a constant issue," they say. "Every time you have to use the restroom in a public place, and you're only offered these two options, it's always a question of, which one do I use? Which one will be a safer experience for me, both from the outside and the inside—of my mental health? Will there be cognitive dissonance? Will I be assaulted? Do I pass for male? Do I pass for female? Every single time you're faced with that, it's an extremely stressful moment."

Just the previous year, the issue had gotten even tenser for trans and gender-nonconforming people, when the "bathroom wars" erupted across the United States, starting in North Carolina. There the General Assembly passed the Public Facilities Privacy & Security Act, which required that people use restrooms in public buildings, including schools, that corresponded to the sex on their birth certificates, even when that conflicted with their gender identity. In the following years, several state legislatures introduced similar legislation.

Advocacy groups tended to push for the rights of trans people to choose the appropriate restroom for themselves, as well as for the addition of separate, unisex restrooms. But Sanders, along with lawyer Terry S. Kogan and trans historian and activist Susan Stryker, saw the conflict differently, as an opportunity to rethink the public restroom altogether. Under the project name Stalled!, they advocated a "universal" or "inclusive" solution: a single, multiuser public restroom that would serve the needs of everyone, regardless of gender identification. Listening to the lecture, Choe was floored. Stalled! was "treating the issue of access to restrooms as a design problem"—one with a potential resolution—instead of as an intractable "political headache," a pitched battle to be eternally waged in statehouses, courts, and the media. By the summer, Choe was working on the project and now they manage it under Sanders's supervision.

Like so many aspects of today's toilets, sex-segregated public restrooms first took hold in Victorian times. At the beginning of the nineteenth century, according to Kogan, men and women both worked out of the home, making, selling, or practicing professions. But in the following decades, the industrial revolution restructured society in a way that divided it into gendered "spheres"—women guarded the private sphere at home, while men ventured out into the public sphere in workplaces, especially factories. To reinforce this, a "cult of true womanhood" arose, which put women on a moral pedestal in contrast to "vulgar" men. Over the century, a (largely bogus) scientific consensus arose to support this cult, establishing that women were intrinsically different from men, not only in their bodies but also in their intellectual capacities and temperaments.

Many women needed, or wanted, to escape the gilded cage of the home, though. In response, Kogan writes, architects and urban planners created women-only sections in the public sphere, ones that would reflect the home space, in order to protect the weaker sex and their potential offspring. These included ladies' reading rooms in libraries, ladies' railroad cars, ladies' parlors, ladies' post office windows, and, of course, ladies' restrooms. Ultimately, Kogan writes, Victorians' convictions about the vulnerability of women's bodies in the public sphere, plus their uncompromising views about women's modesty, led them to conflate the requirement for separate toilets for women "with other requirements of sanitary science related to piping, water supply, or sewage." Unisex toilets became, by this definition, unsanitary toilets. Politicians legislated this separation into law, and government bureaucrats wrote it into building codes.

While this paradigm protects women in some ways, the fallout is hiding in plain sight. Requiring separate toilets for women means that there's an easy way to make women feel unwelcome: don't give them toilets, or make them walk farther for them. "Until quite recently, the complete absence of women's toilets in certain locations clearly signaled their exclusion from halls of power," writes Judith Plaskow:

> Sandra Day O'Connor found when she joined the Supreme Court in 1981 that, unlike her male colleagues, who had their own restroom, she had to walk down a long hall to the public lavatory. Similarly, there were no women's toilets near the Senate floor until 1992. Women in the House of Representatives finally got a bathroom near the House chamber only in 2011. Institutions such as the Princeton Graduate School, Harvard Law School, and Yale Medical School for many years not only had no restrooms for women but also justified women's exclusion on the grounds that there were no available lavatories.

After a 2015 Democratic presidential debate, Donald Trump disparaged Hillary Clinton for taking a bathroom break that ran a little longer than the commercial break. "I know where she went. It's disgusting. I don't want to talk about it," he said. It later came out that the women's restroom was farther from the stage than the men's.

Likewise, while the women's room might serve as an escape from men, as well as a place to share female fellowship, some have noted that the men's room serves as a kind of power center from which women are excluded. "One senior male litigation partner at the major New York firm at which I used to work was notorious for beginning business conversations with male subordinates with the invitation, 'Come pee with me,'" writes Mary Anne Case of the University of Chicago Law School.

Victorian ideas about sex separation also gave us long lines for women's restrooms. Kogan cites the following guidance from 1910: "Ordinarily it will be found that one water closet and one urinal for each 20 male employees, or part of that number, and one more closet for each 20 female employees, or part of that number, is the smallest possible allowance." Did you get that? That's two facilities for twenty men and one facility for twenty women, although one study has shown that it takes about twice as long for women to urinate, from entering to exiting the toilet. Some seventy years following that, "potty parity" laws started mandating that institutions equalize waiting time for women by providing more toilets for them, though men didn't like the ramifications: in 2003, a newly renovated Soldier Field in Chicago converted five women's restrooms to men's restrooms after protests about waits; thereafter, women had to wait twice as long as men on average.

Current restrooms are even worse for people who don't have a clear place in the sex-segregated paradigm, whether by their own judgment or that of others. This includes transgender people. Gavin Grimm, a Virginia high-school student, happily used the boys' room until his high school forbade it; ultimately, he won a high-profile court case. It also includes those who aren't transgender but aren't gender conforming, either. Khadijah Farmer, an African American lesbian describing herself as "not the most feminine," got kicked out of a New York City restaurant and bar in 2007 because a patron perceived her as a man entering the women's room. And then there are people who identify as neither male nor female, including some nonbinary transgender people, as well as some intersex people who have been born with physical characteristics that don't fit standard definitions of male and female bodies.

For many, the collateral damage of sex-segregated toilets has become too much. But what's the alternative? Allowing people to use the toilets consistent with their gender identity and expression—the opposite of the bathroom bills—is an incomplete solution at best. Some argue that men would abuse these rules to enter women's restrooms for nefarious purposes, making women less safe, though the rebuttal is that men who want to enter women's restrooms for nefarious purposes would probably do so anyway.

Adding single-use "gender-neutral" toilets in addition to traditional, sex-segregated restrooms—a common solution today—also isn't adequate, Choe says. "This continues to isolate and segregate and 'other' people who are deemed as different." (People with disabilities also often face this type of restroom segregation.) Recently, Choe got kicked out of a single-sex restroom with a sauna in a gym after another client complained. An employee pointed them to "the single-user restrooms that had showers inside them that were used by children during their summer camps, which were kind of offset from the locker rooms. And of course, those didn't have saunas in them," Choe says.

And Stalled! doesn't usually recommend slapping an "All Gender" sign on existing multiuser restrooms, because the stalls in American restrooms aren't designed to provide the privacy that people would quite reasonably expect. Hanging a curtain to partition off urinals, as some have done, "is, one, kind of gross, but, two, a strange temporary measure that I don't think would really encourage women and non-male people to go in there," Choe says.

So what does the Stalled! collaboration suggest? There's no one-size-fits-all model, Choe says, but a set of general principles, which can be illustrated by a design the team proposed for a sports facility at Gallaudet University: Instead of two restrooms, there's one all-gender public restroom, open to the hallway, with gap-free floor-to-ceiling partitions around each stall and a communal washing-up area. In addition to standard compartments, there are some that are accessible to people with various types of disabilities, as well as caregiving rooms that have sinks and changing tables. Along the corridor side of the toilet stalls, a lounge with seats and tables creates a social space. Instead of signs representing gender, symbols, such as toilets and faucets, represent what can be done in the restroom.

A public-school system in St. Paul, Minnesota, has built something like this. Designed by TKDA Architects, the first prototypes were installed at Johnson High School, an aerospace and engineering magnet school, in 2016. Located on a corner where two hallways meet, their placement allows anyone to look directly into the area; what they will see are handwashing stations and doors to the stalls. Each stall is private, enclosed by floor-to-ceiling doors and walls, and equipped with an automatic fan and light. Locked doors read "Occupied" on the outside and "Secure" on the inside. There are no urinals. The openness and visibility of these spaces, while unusual for restrooms, provide "passive security," discouraging bad behavior such as fighting and bullying—which, in high school, is not something that happens to just transgender students.

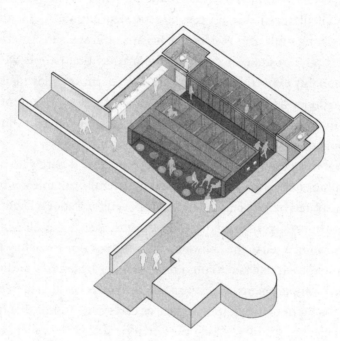

Stalled! prototype design for restrooms in a sports facility at Gallaudet University.

One barrier to making these changes is financial: inclusive models cost about $12,500 more per compartment than typical restrooms, according to a white paper by the architecture firm Cuningham Group, based in part on the St. Paul schools. Another is cultural. For some, universal restrooms are an attempt at an end-run around the bathroom wars. They hope they can win over some conservatives who won't specifically accept transgender rights. That seems to have been the case in St. Paul. The Minnesota Family Council, a Minneapolis-based Christian organization that objects to transgender students in sex-segregated restrooms, told the local paper that they did not mind the all-inclusive restrooms "because transgender students would not be in the same space as their peers behind closed doors."

But people will fight to keep their restrooms as they are: during World War II, a group of white workers at the Western Electric Company in Baltimore went on strike when the company banned segregated restrooms, forcing the military to take over some plants. The company dropped the new integration policy to prevent further disruptions to the war effort.

And even those in the broad coalition of people who oppose the recent bathroom bills—which seem to have fizzled out, at least for now—might find inclusive restrooms to be strange, and at times unpleasant, new territory. Men won't enjoy the experience of waiting in lines for stalls, while many women, who feel pressure to hide their bodily functions, might feel even shyer about using all-gender public toilets than they already do about single-sex ones. But I think people would quickly get used to it: I regularly come across restrooms like this in Europe, and although I still sometimes do a double take when a man walks up next to me to wash his hands, neither his presence nor my double take does either of us any harm. And, in reality, inclusive restrooms—with their private stalls and indifference to gender—are much like what we have in our homes.

For now, the biggest hurdle may be regulatory, since building codes around the country—and, indeed, many parts of the world—still mandate single-sex restrooms. Some organizations seek exceptions, such as LGBTQ Jewish Congregation Beit Simchat Torah, which obtained a special ordinance from the NYC Department of Buildings to create an all-gender restroom. But all that is about to change. In 2019, Stalled!, together with the National Center for Transgender Equality and the

American Institute of Architects, helped push through amendments to the International Plumbing Code that allow for all-gender multi-user restrooms in addition to single-sex restrooms. It was a slow and bureaucratic process, but this change, as it trickles down to local ordinances, means that architects won't "have that excuse anymore of saying, oh, the code doesn't allow it," Choe says. It "will be felt in generations after."

Potty Talk

Let's analyze the shit out of this.

Those who write on shithouse walls
Roll their shit in little balls
Those who read those words of wit
Eat those balls of little shit

—"Traditional" restroom graffiti

Cutting the Crap

In Surabaya, Indonesia, I walk the plank. It's a narrow, rickety wooden board, leading from a modest back garden to a small shed on stilts over a wide, rushing river. The floor inside the shed has a hole in it: it's what the locals call a helicopter toilet because it hovers over the water. A dozen or so people chuckle while I check it out; even though I don't plan to use it, they find it amusing to see a white foreigner, a rare sight in this neighborhood anyway, navigate this typically Indonesian loo. But there's little to laugh at, really: many millions of people throughout this country use unsafe toilets every day, sending their poop on a vile river cruise that has the potential to dock anywhere downstream.

Fortunately, this particular helicopter toilet is about to go out of service. Nearby, two laborers, in shorts and barefoot, hack at the ground, digging the hole where a new simple septic tank, connected to an indoor toilet, will go. The home consists of two basic rooms with a concrete floor, exposed wooden rafters, and few frills. Ratna Ayuningtyas lives there with her husband and ponytailed two-year-old daughter. The young mother moved to the community when she got married. Her previous home was not near a river, so when she started using the helicopter toilet, "I felt embarrassed because I was not used to going to the toilet outdoors. But then the more you do it the more you get used to it," she says. She and her husband bought the septic tank, though they found it expensive, so that their daughter can have a healthier life, she says.

Nearby is the man who sold it to her. His full name is Koen Irianto Uripan, but according to Indonesian custom, he's respectfully called Pak Koen, and he jokes that he's widely known as Irianto WC, for water closet, since that's how people enter his name into their phone contacts. From humble origins, he had been working as an instructor on sales techniques when, in 2008, he happened to fill in at the last minute for a World Bank program that was training "sanitation entrepreneurs," businesspeople who would sell toilet supplies to Indonesia's poor, as part of the country's widespread efforts to improve sanitation. Convinced that the market was ripe, he decided to start his own business.

It's not easy. According to 2017 numbers, some 25.5 million of the country's 264 million people practice open defecation, putting it near the top of the world's rankings. On Java, which is the world's most populous island, people traditionally have an accepting attitude toward the practice and also often use helicopter toilets over rivers or indoor toilets attached to PVC pipes that discharge directly to a waterway. Some hold the belief that defecation into the river is fine because "the flow of the river cleans out the feces." Others claim that the feces feed the fish or fertilize paddy fields. Even if people are not happy with the situation, they may not have the knowledge to fix it: what constitutes safe sanitation, where to buy products, and how to get loans to pay for upgrades. It doesn't help that the country has almost no sewers, even in the densest areas, and that many septic tanks on the market are not watertight. The law leaves it up to every one of millions of households to make sure it manages its waste safely.

Many would-be sanitation entrepreneurs have given up in the face of these challenges, but Pak Koen has found that he has a knack for the job, with great rewards for himself and his customers. He has installed thousands of septic tanks where once there were none. He bought his first car while still living in a low-cost boardinghouse, and now he cruises in an extra-large jet-black SUV. And, while he has made money, he has also helped others, a mix that brings his life meaning. "I see this as being of service to people," he says. "I'm giving people education; I'm giving them knowledge. But at the same time, as a businessman, there's also a business opportunity."

I think of Pak Koen as a sort of modern-day Indonesian Thomas Crapper to the poor—and I mean that as a compliment. You're probably familiar with Crapper's name, but perhaps for the wrong reasons. That's due to a 1969 satirical book by Wallace Reyburn, titled *Flushed with Pride: The Story of Thomas Crapper*, which invented a fictional life for a real man, baptized in South Yorkshire in 1836, whose name must have amused the author. The book claimed (falsely) that, at eleven years old, young Thomas Crapper walked to London from the north of England to become a plumber, and ultimately invented the flush toilet. Although Mr. Crapper was a sanitary engineer and took out nine patents for his own minor inventions, the flush toilet had been invented well before young Thomas's birth.

Perhaps even more surprisingly, we also don't use the verb *to crap* because of Crapper. Instead, his may be one of the world's most famous cases of nominative determinism: picking one's profession based on one's name. The word *crap* as a synonym for *shit* predated his fame, with roots in the medieval Latin word *crappa*, for "chaff." So did the term *crapping ken*, referring to an outhouse or water closet, which circulated when Thomas was still a child.

Even though Crapper didn't invent the toilet, he did market the shit out of it. You might think that the delicate Victorians would have been keen to install more hygienic flush toilets once they were invented. But they weren't, according to Robert Hume, a Crapper biographer. It was an expensive investment, even for the wealthy, and if the pipes leaked— which they often did—they risked damaging ceilings and furniture. Plus, many people considered it disgusting to shit inside the home instead of in the privy outside. (Chamber pots were mainly for urine.)

But Crapper set out to win hearts and minds and to do so came up with a wide range of sales techniques. In 1866, he opened the world's first bathroom-fittings showroom in a posh part of London. There he "established a reputation for sanitary ware of the highest quality," according to toilet historian David Eveleigh. Many of the toilets in the showroom were operational, so customers could see the latest technological developments that made toilets more hygienic and less stinky, a big improvement over the scaled-down sales samples commonly presented to customers at the time. Crapper also put toilets in his shop window, which apparently was shocking enough to cause some ladies to faint.

Around the world, major changes in the way we handle and manage human waste are poised to take place if people will embrace them. What it takes for Pak Koen to change his customers' minds about helicopter toilets can't be so different from what it would take to change other people's minds about the health, environmental, and social aspects of their toilet systems. And it's not just about changing people's minds; it's also about changing their priorities, their behaviors, and their spending habits.

Though it comes with some ribbing, there can be glory for those who lead on the question of toilets. While Crapper didn't originate the word *crap*, he may be the reason we use the term *crapper* to refer to the john. Hume posits that U.S. soldiers based in England in World War I saw "T. Crapper" on toilet cisterns, found it funny, and brought the word home to America with them. We don't know what Crapper would have thought of this form of immortality, since, in 1910, he died, sadly, but ever on brand, of bowel cancer.

Poo and Taboo

In 2013, actor Matt Damon, co-founder of the nonprofit Water.org, claimed in a staged press conference that he would be going on a toilet strike until all people have clean water and sanitation. It was, of course, just a bit of faked news. The goal was to get people thinking. But why do we always have to make a joke out of toilets? Why can't we just talk about them with a straight face?

When it comes to toilets, our psychological baggage is heavy. One of the items we have packed away is disgust. As noted earlier, disgust can act as a behavioral extension of the immune system, keeping people away from dangerous substances like rotten food and poop. But, like the body's immune system, disgust can also overreact, creating taboos. In the 1980s, psychologist Paul Rozin performed an experiment in which he showed that subjects wouldn't eat fudge shaped like a dog turd, suggesting that anything connected to poop, no matter how nice, is tainted by association. Disgust's effects also extend into the social realm. One line of research, initiated by Rozin and Jonathan Haidt, has advanced the theory that the emotion of disgust evolved into a response to moral violations. Some studies in this area tend to proceed like this: researchers expose subjects to a disgusting taste, smell, sound, or image—dirty toilets are popular—and then ask them what they think of morally transgressive acts unrelated to the stimulus, such as first cousin marriage. In quite a few studies, the subjects expressed more condemnation when they'd first been disgusted by the stimulus. That is, grossed-out people might be more judgmental. Although this theory is undergoing reevaluation, it could present a particular problem for the toilet revolution: just thinking about toilets might induce disgust, making people less open-minded to alternatives and perhaps even more prone to seeing them as immoral, with reference to everything from the reuse of water to the reconfiguring of gendered labels.

Studies have found an association between a high sensitivity to disgust and an authoritarian attitude, which includes prejudice.

According to other research, the more that people fear contagion, the more prejudiced against outsiders they seem to be. And studies have shown that people who live in countries with a higher threat of parasitic infections tend to be less open-minded to ideas and experiences—and that those countries are less likely to have robust democracies, individual freedoms, and economic and gender equality. It's tempting to see toilets as a secret key to democracy and liberalism.

On a lighter note, the emotion of disgust may even drive the long-standing marital conflict around whether men should put the seat down after they urinate, so that women will find it in the appropriate position, risking neither the humiliation of falling into the toilet nor contamination from touching the seat. Nick Haslam, charmingly, devotes an entire chapter of his book *Psychology in the Bathroom* to this question. Although the most efficient arrangement would be for everyone to move the seat to where they need it when they need it, he argues that a cold economic analysis is not the right approach to the problem. Some women, he observes, see a raised toilet seat as "an in-your-face display of male privilege, a blatant insistence on male difference." And women tend to be more disgust-sensitive than men, making touching the seat more unpleasant for the fairer sex. "Overall, the burden of disgust falls on women more than men. They are held to a higher standard of purity and cleanliness, sanctioned and humiliated more for violating this standard and expected to distance themselves more from the products of their bodies or even to pretend that there are none," Haslam writes, exhorting men to just put the seat down. "Sparing women unnecessary disgust should not be too much to ask of men."

As commendable as I find Haslam's conclusion, I have a preferred alternative: however onerous the task may be, all people ought to put the *lid* down when they are done, as that is what actually prevents germs from flying around the room when the toilet is flushed.

We consider dwelling on excretion to be a sign of a disturbed mind. This goes back at least to Sigmund Freud, the father of psychoanalytic theory, who believed that human development was marked by an "anal stage," in which children experienced pleasure in learning to control their bowels and saw their poop as a valuable commodity that they could offer as a gift. Since Freud associated this stage with the development of self-control and will, if children failed to successfully pass through it, perhaps because of harsh toilet training, they could develop an "anal" personality, characterized by the "anal triad" of orderliness, obstinacy, and parsimony. (Interestingly, it is in this context that Freud himself quotes the aphorism, later interpreted by anthropologist Mary Douglas in her groundbreaking work, *Purity and Danger*, that "dirt is matter in the *wrong* place.")

In the first decades of the twentieth century, Freud and his successors developed a detailed picture of the anal personality. As Haslam notes, they applied it to figures as divergent as Goldilocks' three bears (fastidious beasts who despise having the girl, whose golden hair color, bizarrely, symbolizes feces, mess with their stuff and put her rear on their chairs) and Richard Nixon (an obsessive hoarder of intelligence who lived in terror of leaks). Psychoanalysts also interpreted constipation as a hidden desire to possess and retain; diarrhea, as a secret desire to give or eliminate. In one case study, a young married woman whose husband wouldn't give her the baby she desperately desired went two years without a spontaneous, unaided bowel movement, until one day he brought her a bouquet of flowers for the first time.

Current psychology has more or less discarded Freudian doctrine, but Haslam argues that Freud and his successors' description of the anal character type had some merit—although the connection to toilet training does not seem to exist, thank goodness. It has essentially morphed into the contemporary description of obsessive-compulsive personality disorder (not to be confused with obsessive-compulsive disorder—a separate, better-known diagnosis that involves ritualistic behaviors). Traits associated with the personality disorder include perfectionism and conscientiousness, as well as, perhaps, an especially acute aversion to the gross. Growing up in the 1990s, my cohort lived with regrettable residues of Freudian theory. It was popular, at least in my circle of tweens and teens, to describe supposedly uptight, withholding people as "anal" and "anal-retentive." I lived in fear of the label being linked to me.

Urination never got the same attention from psychoanalysts as defecation, but it's also in our toilet-related baggage. Freud thought humanity took an important step toward civilization when our ancestors overcame the "homosexually tinged" urge to pee on fire. That's hard to prove, and actual sexual fetishes with urine, such as "golden showers," are apparently rarer than popular media reports would suggest, but what's real is that some people—quite a few—suffer from paruresis, what's known colloquially as "shy bladder" or "bashful bladder," which makes it difficult or impossible to pee in a public restroom, often for fear of being seen and judged. Childhood bed-wetting, while embarrassing to the child, is usually biological, and not psychological, in origin and tends to resolve with age. But that didn't stop psychologists from floating a (now-discredited) theory that a triad of childhood behaviors—bed-wetting, fire setting, and animal cruelty—predict criminality and violence in adulthood. For our children's mental well-being, we would do well to call these ideas out for the bullshit that they are. Defecation and urination are normal and healthy—and so is the urge to contemplate them.

Because shitting is such a psychological minefield, we don't just refuse to talk about it, we also hide the act—and the technology we employ for it—from our own internal thoughts. Scientists refer to "attentional avoidance," the impulse to turn one's face and thoughts away from certain things. Just try to count the number of times you used a toilet today; then, as a comparison, try to count the number of times that you ate food or drank coffee. Or take another example: After I told an acquaintance of my interest in toilets, she complained that hers was off-kilter and leaky. When I asked her whether she'd told her landlord about it, she admitted she hadn't. Although she bitterly cursed the situation while in the bathroom, upon leaving she would immediately forget about it.

This tendency to "flush and forget" (or plop and forget, in the absence of a flush toilet), if not completely universal, is widespread. The historical record, from most cultures, contains precious few mentions of the toilet. And studies of swearing show, in a kind of backdoor way, that the poop taboo connects us all: one researcher, in analyzing twenty-five languages, found that "faecalia," or poop-related swear words, were the most common, beating out those with connections to religion (*damn*), animals (*bitch*), slurs (not gonna put one here), and other themes. In English, *shit* is in the top two of most frequently said swears, which serves to illustrate its power, as well as its versatility.

This has consequences for our infrastructure, says Patrick Moriarty, the CEO of the nonprofit water and sanitation think tank IRC in the Netherlands, since we also become blind to the systems that keep our toilets working. "We are not just talking about pipes or pumps or toilets; we're talking about the people and the rules and the frameworks and the institutions that go around all of that hardware to make it deliver a service." That encompasses a vast network of actors, including public utilities, regulators, businesses, banks that lend money to the businesses, tax agencies, and even the schools that educate the people who work in those places. Most of us don't ask—until our produce gets recalled—whether the people who pick our vegetables should really have to be squatting in the fields to relieve themselves. We don't worry about how people who are incontinent manage their toilet needs when out of the house—until we find ourselves in desperate need of a public toilet ourselves. We don't contemplate the resilience of our wastewater treatment plants until a storm overwhelms them, coating our cities in fecal matter. And we don't think to ask why the toilet hasn't changed in the past 150 years even as we absorb, adapt to, and sometimes fight transformative technological change in almost every other segment of our lives.

I have hope that there are ways to overcome these hang-ups. Sociologist Harvey Molotch writes that "to deal with waste intelligently, it has to become . . . approachable." I'm thinking of the Solrødgård Climate and Environment Park in Denmark, whose architects combined a wastewater treatment plant with green-roofed buildings, a recycling center, walking trails, a birdwatching tower, and a bat hotel. Or the Tokyo Toilet project in Japan, which commissioned leading designers to come up with bold restroom ideas for public parks—including cubicles with transparent, jewel-toned "smart glass" walls that allow for safety and cleanliness inspection from the outside but turn opaque once occupied.

Other advocates of improved toilet systems draw people in by linking sanitation to higher-profile topics such as human rights, the economy, and health. Slum Dwellers International, a global social movement of the urban poor, has made toilets a symbol in its fight for slum improvement. And, recently, environmental justice activist Catherine Coleman Flowers, founder of the Center for Rural Enterprise and Environmental Justice and known as the Erin Brockovich of Sewage, has managed to push rural toilet infrastructure up the U.S. political agenda. Her grassroots strategies include partnering with scientists to test soils for parasites, inviting high-profile politicians to witness the dire conditions in which some Americans live, and connecting the problem to climate change, which she calls the greatest challenge that humanity has ever faced.

Another way to make toilets approachable might be through potty humor itself—what Haslam, when discussing funny farts, calls "a controlled violation of social norms of propriety." Crapper knew that he could use the taboo to his advantage, employing an early form of shock marketing. Indeed, many brands have since used the taboo. A twentieth-century example can be found in the high-tech toilet known as the "washlet," which comes with features such as water sprays for the back and front, blow dryers, heated seats, automatic lids, and even musical accompaniment. The washlet originated in the United States, invented in 1964 for health-care facilities by the American Bidet Company as the Air Wash Seat. Yet it really took off in Japan, where, in the 1980s, the company Toto put a vastly improved version on sale for the mass market at about the same time people were shifting away from the traditional Japanese squat toilet. Today, washlets can be found in more than three-quarters of Japanese households (as well as in some public restrooms, including the transparent ones from the Tokyo Toilet project). One key to its success there was a famous TV advertisement featuring a demure young actress spouting the shocking slogan "Even though it is a butt, it would like to be washed." Despite years of dedicated effort, however, the company has never been able to re-create the washlet's success in the United States.

Or take an even more recent commercial, which you may have seen: A prim young woman sits on a toilet in a stall, knees squeezed together, billowing skirt barely concealing her underparts. "You would not believe the mother lode I just dropped," she says, with a cheeky grin. "And that's how I like to keep it." Although it's for a spray, called Poo-pourri, meant to conceal bathroom odors, it's not earnestly euphemistic in the way that menstrual product or antiperspirant ads tend to be. Instead, we see the young woman transported on her toilet to an office, an apartment, and a flowering field, using vulgar terms like "pinch a loaf," "cut a rope," and "lay a brick."

Toilet advocates have taken a page from this book, too. One is Jack Sim, who's known as Mr. Toilet or, sometimes, the Hugh Hefner of Toilets. In the late 1990s, Sim was a self-made construction and real-estate entrepreneur whose own fortune had risen with that of his home country, Singapore. In his forties, he was looking for a new purpose, a way to "live a useful life" (not a sentiment I would identify with Hefner, so I think the comparison doesn't go very far). Despite growing up poor, Sim had never considered the problem of inadequate toilets. But then he read a newspaper article about Singapore's dirty public restrooms and saw an opening. His main theory was that "what you don't talk about, you cannot improve."

That meant stimulating lots of conversations about toilets, in any way that worked. In 1998, he founded the Restroom Association of Singapore and then, in 2001, the World Toilet Organization, a global network that connects anyone with an interest in toilets and which convenes an annual World Toilet Summit. He declared November 19 to be World Toilet Day and got the United Nations to recognize it.

He also encouraged experts and entrepreneurs, talked to newspapers, participated in a documentary about himself, and told jokes. Lots and lots of jokes. "This morning I was invited to *BBC Breakfast* live broadcast, and they told me that they will interview me about my work, but on one condition: that I cannot use the word *shit*," he said a few years ago, smiling sweetly, at the start of one talk. "So I complied, but I feel a little bit constipated."

In today's crass social-media cacophony, it might be hard to imagine just how subversive jokes like this once were—and how effective. It would be a mistake to think, however, that humor can do all of the work. If shock and jokes were enough, there would be universal sanitation, not to mention that everyone would have Toto toilets. Plus, there's the real risk that people won't look past the joke and see the serious message beneath. Getting people to change, it seems, requires a deeper plunge.

Game of Thrones

Just as we need innovation in technologies, we need innovation in the way we inspire transformation, whether it's with toilet users, wastewater engineers, or politicians. In Zambia, a landlocked copper-exporting country in southern Africa, researchers from the London School of Hygiene & Tropical Medicine (LSHTM) and the Centre for Infectious Disease Research in Zambia (CIDRZ) set out to see if they could improve the state of sanitation, not by building better toilets or making new laws but by convincing people to make changes themselves. They were applying a method called Behavior Centred Design (BCD), developed by Robert Aunger and Val Curtis at the LSHTM. It aims to "produce changes to the environment, which causes changes in the brains of the target audience, which, in turn, cause them to behave differently."

The toilet situation in the Bauleni slum in the city of Lusaka is, as in so many slums, bleak. The toilets tend to be shared by several families on a single plot, who are renting from the same landlord. Then–graduate student James "Ben" Tidwell of LSHTM, who carried out the project along with Jenala Chipungu in Zambia, described one typical facility like this:

> The toilet's walls were made from a combination of cement bricks and plastic tarp, and the floor was concrete. But there was no roof, and the wooden "door" was not actually attached to the structure. The toilet was poorly ventilated, rarely cleaned, and full of flies. It had barely enough space in which to turn around. The pit, which contained the waste, was simply a hole in the ground, and [one resident] had heard stories of toilets in the area collapsing while in use. The pit overflowed any time there was the slightest rain.

Nonprofits had told landlords that the toilets weren't safe for their tenants, but knowledge often isn't enough to prompt change, even when there's no money involved. A simple example is handwashing: people know that they should wash their hands after using the toilet or before handling food, but they don't always do it, even if they have ready access to water and soap.

First, and importantly, the LSHTM/CIDRZ team did wide-ranging interviews with the people involved, to listen to what they wanted. Tenants told the team that they valued good toilets—so much so that they were willing and able to pay higher rents for them. But the landlords had no idea that they were leaving money on the table. They hadn't spoken to their tenants about it. In fact, the landlords often didn't really know their tenants at all, despite sometimes living on the same plot.

Public health experts are usually not marketing geniuses. So, instead of coming up with a campaign themselves, the team sent a brief about it to a local professional creative agency—the modern Mad Men (and Women) of Zambia, you could say—specifying the changes they wanted: a rotating cleaning system for the tenants, a lock on the inside and the outside of the toilet door, and a sealed toilet that wouldn't let flies in. The team also instructed the agency to use positive instead of negative (shame-based) messaging. The result was a stealth marketing campaign that Curtis once described as following a "how-to-get-rich American model." They called it the Bauleni Secret.

Via a whisper campaign in the summer of 2017, they invited landlords to secret society–style meetings with the promise that they would learn tricks for making more money and reducing conflict on their properties. Trained actors and neighborhood health committee members started out by presenting new knowledge in a fun, attention-grabbing way—short videos of what appear to be real-life scenes captured by a bystander with a smartphone (although they're really staged). The first video shows tenants fighting publicly over who hasn't fulfilled their responsibility to clean the toilet, ultimately revealing that the tenants blame the landlord. The second shows tenants struggling to hold a toilet door closed while they're using it, resulting in a man accidentally walking in on a woman. The third shows some drunk men wandering onto a plot and using a toilet that's not theirs, creating a mess. The final is the most powerful, though: it shows couples visiting a house for rent and turning it down because of the disgusting state of the toilet. "This is basically a cholera center. I'd never bring our kids here," one woman says to her husband as they leave, outside of the landlord's earshot. They decide to return to the first rental they'd seen, where "the rent's a bit much, but at least the toilet's clean."

After watching the videos, the landlords participated in hands-on games and demonstrations that helped them to process the new information. Finally the program helped the landlords act on their new knowledge, through buddy systems and savings programs.

To see if the program worked, the LSHTM/CIDRZ team set up a randomized controlled trial, providing the intervention to half of more than a thousand landlords in the area of sixty-four thousand residents. When the researchers followed up just six months after the start, they found that the plots of the landlords who had gotten the intervention were more likely—by about ten percentage points—to have implemented each of the four improvements than those who didn't and some even did more thorough renovations. Toilets were more hygienic and private. And Tidwell's pretty sure that some of the changes continued to kick in later, especially as word spread through the community. "We saw people that just made simple improvements. We also saw people who built entirely new from-the-ground-up toilets as a result of the intervention," he says. "We'd love to study it over longer timeframes."

Tidwell is careful to say that this kind of approach can't fix the whole problem. Why? One, slum residents don't necessarily fret about international definitions of "safely managed sanitation" like toilet experts do, but they do care about having clean, secure, and pleasant toilets, so that's what they'll pay for. Two, poor people won't be able to afford all the needed toilet improvements by themselves—especially the costs associated with managing the sludge in the pits. And it's not even fair to ask them to do so, since wealthier people often benefit from government subsidies for public sanitation services such as sewers and treatment plants. "Pro-poor policies and subsidies and regulatory environments and product innovations are very much needed to fully address this challenge," Tidwell says. Still, he and his colleagues estimate that if landlords were to make all of the improvements that tenants are willing to pay for, then the percentage of plots with flushing toilets (which people want) would shoot up from a current 15 percent to 58 percent and those with sturdy buildings would go from 42 percent to 72 percent.

And the implications are much greater. The demand for better sanitation services is not limited to slums: in the United States, according to a 2020 poll of more than a thousand voters by the Value of Water Campaign, 70 percent "want the president and Congress to develop a plan to rebuild water infrastructure" (which includes wastewater), 84 percent "support . . . increasing federal investment to rebuild the nation's water infrastructure," and 73 percent "support investing in water infrastructure to increase resilience to climate change," *even with a $1.27 trillion price tag*. If the Bauleni Secret could convince reluctant landlords in Zambia to part with their own money to upgrade toilets, then there's hope that new ways of "talking shit" can recruit reluctant politicians to fund the rebuilding of U.S. water infrastructure. Where shock and shame and well-meaning lectures don't work, there is always another way in.

The Bottom Line

One of the most unlikely encounters I had while doing research on toilets was with a young Japanese designer named So Andrew Saito, who had made himself notorious with his colleagues by firing his own poop in a giant kiln in a ceramics studio in the Netherlands. When I first heard of him from a friend, I was skeptical: an artist obsessed with his own poop is such a cliché, I thought. But when we spoke, I realized that he was not a navel-gazer who had ventured a little south of the belly button, but a curious soul who had fallen down a rabbit hole (or manhole) much like the one I was in. His interest had developed when he realized that he could make ceramic glazes from the ash left over from the incineration of sewage sludge; the ash's mineral content would generate lovely colors (for example, iron makes a light blue glaze under certain conditions). Entranced by how he could incorporate ceramics with waste treatment, he ended up designing an entire circular system for a theoretical rural village, in which people would use kilns to treat poop and poop as fuel to fire kilns.

Unfortunately, the design process took a lot of trial and error—and the incident at the studio was an error. "The smell was terrible," he told me, deeply embarrassed by the misery he had caused his colleagues. "I almost fainted." I met him when he was finishing his project, yet, far from thrilled that he was nearly done, all he could think about was how he could take the next step to making his idea into a reality. I don't know if he will manage, but his work has broken new ground into what's possible with ceramics, waste, and rural livelihoods.

When it comes to our toilet systems, we're on the precipice of a paradigm shift, but we could teeter here for a long time to come if we don't, together, actively decide to make some changes, despite the risks they represent. This isn't a book with a single, specific prescription for action, but if you feel moved to do something that will launch you off the page and into the toilet revolution, here is my suggestion. Pick just one or two of the following actions—whatever intrigues you most—and see where the questions lead you:

- *Learn*. Where does your wastewater go after you flush? If to a septic tank, what kind and how does it work? If to a sewer, where is the treatment plant and how does it work? Where do the liquids and solids end up? Is energy or fertilizer recovered? Go on a tour, and take your friends and family.

- *Upgrade*. Can you switch to a toilet with a lower flow? Can you flush with water you've saved from elsewhere, like the shower? Can you buy more sustainable toilet paper or a bidet? Have you enforced a three *p*'s rule for your home—only pee, poop, and (toilet) paper go down? Do you have a step stool nearby to give yourself and the others who use the toilet an option to "squat"?

- *Investigate*. Does your local sewer system leak or overflow? Does your community have a problem with septic tanks that pollute local waters or soils? Who in your community bears the brunt of this pollution? Can you join a local citizen scientist initiative sampling the local water quality?

- *Advocate*. Are the shared toilets in your community's public places, workplaces, and schools sufficient in number and quality? Are they inclusive? Are period products available and, even better, free? Does anybody in your community lack safe toilet systems at home?

- *Experiment*. Are there any groups in your area, like Vermont's Rich Earth Institute or university laboratories, testing new systems, collecting urine or feces, or selling products? If you want to try fertilizing your garden with your own pee and poop, make sure to know the local regulations and follow guidelines from reliable sources, and then start with a urine collector or compost toilet on your property.

- *Support.* Lots of organizations, some mentioned in these pages but many that I haven't named specifically, are doing good work, nationally and internationally. If you have a particular interest in a place or group of people, can you make a contribution, financial or otherwise, to help them access good sanitation?

- *Talk.* Encourage more—and more positive—conversation about the issues around toilets. Can you bring the topic up at a party? Can you participate in or even organize a World Toilet Day (November 19), Menstrual Hygiene Day (May 28), or World Water Day (March 22) event in your area?

- *Defend.* Take the shame out of toilet issues. Ask yourself: How would you speak up if you heard someone being shamed for peeing, pooping, or menstruating?

- *Appreciate.* Wastewater workers usually only hear from the public when something goes wrong, but like firefighters and EMTs, they help keep you alive. How can you thank them, and back them up when they demand safe conditions and good pay?

- *Train.* Would you like to be part of bringing about the future of toilet systems? A "silver tsunami" is about to hit in the United States, with many wastewater professionals in particular nearing retirement age, and plumbers are always in demand.

- *Create.* Contribute your strengths. Even if you don't want to work in wastewater, think about what you can do. Are you an artist? A computer programmer? A writer? A teacher? A real-estate developer? A social-media influencer? A crafter? A restaurant owner?

We can all bring our skills to bear to make sure that our toilet systems work well and for everyone.

Perhaps you don't see yourself as a do-gooder (or a loo-gooder). That doesn't matter. The fact is that the toilet revolution doesn't require altruistic motives. Selfish ones will do. On our shared planet, if just one person doesn't have a decent toilet, it puts everyone at risk. Disease, pollution, suffering—the fallout of these is not easily contained to the places where they originated. If sanitation doesn't work for all of us, it works for none of us. That's the pain of it, and the beauty. We are all connected to one another and to every organism on the planet through our toilets. We all have to get to Loo-topia together.

Life of a Salesman

Unlike Crapper, Pak Koen can't rely on his customers to come to a showroom. So, for two days one recent September, I shadowed him through Surabaya, a dense metropolis of 10 million in the Indonesian province of East Java, as he traveled along the city's endless winding highways, from neighborhood to neighborhood, municipal building to municipal building, living room to living room.

In the first community we visit, the neighborhood with the helicopter toilet, two noisy car bridges bookend a straight and narrow lane, which is lined with small houses and paved with zigzagging bricks painted blue, yellow, pink, and green, the result of a beautification initiative. Residents work as seamstresses, contractors, forklift operators, hatters, and homemakers. Some are unemployed or retired. Almost every front porch and stoop is occupied; songbirds tweet from cages.

Some friendly public health officers in white and beige uniforms meet us. They are a sign that Pak Koen's business operates as part of a larger system. In the mid-2000s, Indonesia began rolling out a new approach to toilets based on the method of Community-Led Total Sanitation (CLTS), which aimed to ignite communities—especially rural ones—to end open defecation without offering subsidies. Between

2006 and 2011, the $14 million World Bank–led Total Sanitation and Sanitation Marketing (TSSM) program in East Java, which gave Pak Koen his start, verified 2,200 of 6,250 participating communities as "open-defecation-free," built 1.4 million toilets, and reduced diarrhea prevalence by 30 percent. It was a big achievement, but imperfect, and many pockets of open defecation and other forms of unsafe sanitation remain, especially among the poorest.

It has taken Pak Koen several presentations over two years to get most households in this neighborhood to buy his product, the simple septic tank connected to an indoor toilet. But there are still a few who have not. So, today, Pak Koen speaks again in a resident's living room. There are more than a dozen community members here, including local leaders and health officers, sitting in a circle on the rug-covered tile floor, forgoing the plush sofas against the wall. It's a social event, and Pak Koen has bought plates of food—peanuts, bread, fried tofu, and bananas—as well as small bottles of water, which sit in the center of the circle. The attendees are mostly female: here women tend to have the final say over matters of household hygiene, though they usually have to get approval from the male heads of household for big expenditures like a septic tank. Most of the women wear scarves that cover all of their hair and wrap tightly under the neck. Two cuddle babies.

Aside from the food, Pak Koen is the main attraction. Born in the East Javan city of Malang in 1962 to a retired military officer and his wife, he was the fourth of six siblings. Money was tight, and his parents eventually returned to the village of their birth, which was almost a hundred miles northwest, leaving Koen with a series of friends in the city. In the house of a former colleague of his father's, he was responsible for tending the chickens. He stayed with his parents in the village for stretches of time and describes the toilet that the family used as a simple pit surrounded by bamboo walls with holes in them and no roof.

Koen earned a place in higher education but decided to go straight to work after graduating. After stints in government and in a factory, he found his calling, in 1986, as a salesperson for the Swedish company Electrolux in Surabaya. He started by selling vacuum cleaners door to door but ultimately got promoted to branch manager—a rare achievement for someone with just a vocational high-school diploma—and then got hired as an executive in the Indonesian branch of the Dutch water-heater company Daalderop. In 1998, he decided to start his own business as a trainer and motivational speaker for sales; ten years later, he discovered the toilet business. In the meantime, he has gone back to school for advanced degrees in law and management and is now working on a doctorate (earning him a new moniker, which translates as Dr. Toilet).

Today, standing in the front of the room, his customers sitting at his feet like pupils, he cuts a distinguished figure—slim with receding gray hair, a neat mustache, and rimless glasses, his phone stashed in the breast pocket of his short-sleeve button-down shirt, which is open at the neck in the local style. For this audience, he speaks in a mix of Bahasa Indonesia, the official language of the country, and the local dialect of Javanese; even without my translator I can get the gist, in terms of not only the information he's sharing but also the tone he takes. He's straightforward and no-nonsense but also hilarious, sometimes silly, and kind, like a favorite uncle. He doesn't condescend to, judge, or patronize his clients, and they reward him for it with their rapt attention. Most of them have already bought septic tanks but are here to take part in the fun.

Pak Koen explains the system itself, in detail that goes well beyond what I suspect most people anywhere know about their toilets—evidence, to me, that he holds his customers in high regard. Users squat over a ceramic pan and then pour water in to wash the waste down a pipe into a watertight fiberglass tank, which he helped design. It's about the height and circumference of an adult human holding out his arms. There the solids settle to the bottom, while oils and fats float to the top. The liquid in the middle then discharges to a shallower tank, from which it drains into the soil. The aboveground footprint is two cement circles and a tall pipe that releases gases so they don't build up inside, which would risk explosions. Households need to avoid harsh cleaning supplies in the toilets, he says, to avoid killing the bacteria in the septic tank that break down the toilet waste. (Even when the system is working well, however, users will have to hire a separate service to empty residue from the tank in about six to eight years, and the fate of that sludge remains a problem to be solved in much of the country.)

At first, I'm surprised that Pak Koen doesn't offer more selection—after all, he could make more money if he sold more luxurious options—but the simplicity is the point. Years ago, locals would have to take multiple trips to various suppliers and hire laborers for a full seven days to install a system that might or might not have been safe. Pak Koen's laborers can do it in a day and a half, with no hassle to the buyer beyond providing a few meals to the workers. Every decision a household has to make can slow its progress down, so Pak Koen offers a "one-stop shop" for a single, reliable product. You say either yes or no.

Pak Koen showing his papier-mâché poop and
holding a box of food during a sales presentation.

When Pak Koen first started training sanitation entrepreneurs for the TSSM program more than a decade ago—before he became one himself—he wouldn't even utter the words for *toilet, latrine, septic tank*, and *poop*. Today, however, he revels in making people blush with crude words and jokes—and they seem to love it. Reaching into his black backpack, he pulls out a piece of fake poop, which he's made out of papier-mâché. The audience gasps and giggles. Then he picks up some food and brings it and the poop in direct contact. Everyone laughs, somewhat uncomfortably. Then he offers the poop-touched food around, and people's faces scrunch up with part-real, part-feigned disgust. At one point, he tries to hand the papier-mâché poop to a woman in a black headscarf, who refuses to take it, looking embarrassed. This woman, it turns out, is one of the people who haven't bought a septic tank yet. Seeing the shocked look on her face, he moves on, satisfied that the message got through: she is making everyone else touch her feces.

Like other programs built on CLTS, the TSSM program came under fire for shaming. One unfortunate advertising campaign featured a disfigured old cartoon man named Lik Telek (Uncle Feces), who openly defecated, spied on women, and came off as generally disgusting. Captions in the advertisements read: "Open defecation spreads disease and stench; shameful, isn't it?" and "My village is clean and healthy. No stench, no flies, and no more Lik Telek. The whole village is more dignified." As a result of that campaign, according to a study by researchers Susan Engel and Anggun Susilo, at least one poor man felt that he had been identified as a Lik Telek by the village head and his staff, who teased him, saying, "You have a gorgeous wife, if you are not making a proper toilet, we may see your wife." Pak Koen's touch is far lighter—to his mind, he's showing people the reality of their situation so that they can do something about it. If they feel shame, it's a by-product of his engagement with his customers, not his intention.

Next, Pak Koen gets to the question of money. He explains the costs. His customers pay him in six installments of about twenty dollars. (The average income in Indonesia is around three hundred dollars per person per month.) So that he can front the money, he tells me, he takes personal loans that he guarantees with his own property, including his house. He started doing that because he couldn't find a bank that would lend directly to people who not only are poor but also have limited collateral. Even though his customers sometimes pay late, their word is on the whole very good, especially since he has personal relationships with their neighbors and community leaders, who can pressure them to pay their debts, a matter of honor in Indonesia. Local organizations, like doctors' professional associations, might provide some funding to help cover part of the costs.

Over the course of the meeting, Pak Koen fields several questions about a new law that prohibits discharging toilet waste directly into the river and whether the government might issue fines for violating it. That has brought some needed urgency to his sales pitches, he says, because sometimes his customers have other priorities. People everywhere spend money in ways that align with their values. As part of the TSSM program, one study aimed to figure out where toilets would figure among residents' priorities in villages in East Java. The researchers surveyed people with no toilets, unsafe toilets, and safe toilets, as well as with professions such as farming, carpentry, teaching, and factory work. Any home renovation, including toilets, the study concluded, came near the bottom of the list, after food, rent, and other routine expenses; gifts for special occasions like weddings and deaths; children's education; debt repayment; savings in the form of items that can be resold later (like gold jewelry or a goat); and buying a luxury good that will raise quality of life (like a TV or fridge). I go into one relatively large home that exemplifies this: they don't have a septic system, but they do have notable status symbols—a professional family portrait printed in a large format, a quality refrigerator, a washing machine. They even have a nice indoor toilet that drains into the river, and would like to upgrade to a more luxurious sitting toilet instead of a squatting one, to save their aging knees. They have long planned to install a septic tank as part of a larger renovation but haven't gotten around to it yet.

It's time to close the deal. Pak Koen tells a dramatic story of one family so committed to upgrading their toilet that they installed a septic tank in the center of their ten-by-eleven-foot living room, since they had no garden. His message: if they can do it, you definitely can—no excuses. "I need only a photocopy of your ID card, and then this whole process can start," Pak Koen says. And, at least this time, it works. The remaining holdouts hand over their IDs. Everyone's happy, since this neighborhood can finally be declared open-defecation-free. Photos are taken; an article is written for the local paper, the *Jawa Pos*, which has played a role in holding the government accountable for progress in sanitation. That's when I realize that I've inadvertently become part of his pitch, too: the next day, the newspaper features a photo of a slightly jetlagged, extremely sweaty me.

My second day in Surabaya offers a less happy story. I follow Pak Koen into a different corner of the city, which is inhabited by waste pickers. The four households on this property—a total of about sixteen people—have come from a different part of the island to eke out a living from the detritus of Surabaya, and make their home among enormous bundles of recyclables: plastic bottles, old paper and cardboard, rubber, and rags, which they collect and sell.

There's not a toilet anywhere to be seen, much less a safe one. I watch a woman wash her clothes in a bucket of water drawn up from a shallow well, and I learn that she and her family also defecate in that same roofless enclosure, letting their feces run down the drain. Other families dispose of their feces down PVC pipes that jut up out of concrete slabs. Most of the waste gets washed by underground flows into the nearby river, but one family here has no running water underneath the pipe. In that case, their poop just collects in an unlined hole they've dug under the slab. The pipe is uncovered, allowing critters to move freely between the underground waste and the aboveground world, further contaminating the space where they live. One public health officer who is there with us says that she often sees diseases related to sanitation among waste-picker families, including intestinal, skin, and respiratory infections. When she first sees my translator and me, her eyes get wide and she runs inside to grab a box of surgical masks, worried that the stench will overwhelm us. (Thankfully, we are more robust than she imagines.)

The government will subsidize much of the cost of two communal septic tanks, but a landlord-tenant conflict is getting in the way. (This chapter's third such conflict, if you're counting—a sign that the right to sanitation is tied to those of fair housing and land ownership.) It's not clear whether the waste pickers or their landlord should be responsible for the small co-payments that the government won't cover. One old woman, who lives in a tiny dark room with two other elderly waste pickers, tells me that she barely makes enough to eat, so she couldn't possibly come up with money to pay for a septic tank. "My life is difficult enough," she says. "Sometimes I cry if I don't collect enough waste to sell to my boss." I'm not inclined to doubt her, but a health officer points out that she wears jewelry and that the waste pickers' extended families live in nice houses thanks to the money they send back. The health officer thinks that they could divert a little bit to a toilet if they want to, but that they have other priorities—jewelry has resale value, after all, and family back home is important. An outspoken woman says bluntly that they don't want toilets because it's just another thing to maintain. Pak Koen can't seem to establish the same rapport with this group that he did with those in the previous community.

Soon, the landlord shows up to discuss the problem. He is uncomfortable, crossing his arms, standing to the side, refusing to sit down. He can't commit to paying for anything, he says, because he co-owns the property with other members of his large family, whom he needs to consult before spending any money. Plus, he rents the waste pickers the land, and the structures on it are technically their responsibility.

The parties are at a frosty impasse; nothing will happen today—and the only people who really seem to care are Pak Koen and the health workers. Frustrated, Pak Koen makes a decision: if they won't agree to the co-payments for a septic tank, he will put his own money toward installing a toilet pan to seal up the most dangerous open pipe, the one that's over the hole in the ground. That will make it marginally safer, provide a measure of dignity, and help create the habit of using a toilet. And he'll come back another time, hoping for a warmer reception.

Later, over a cold drink in a juice bar, we shake off the tension of the morning, returning to the story of his successful career in sanitation. What's the secret to selling toilets in the face of indifference, rejection, and sometimes outright hostility? I ask him. He ponders for a moment. Any business is about relationships, he replies, whether it's pitching electronics to the middle class or septic tanks to the poor. "First, I have to sell myself. I have to make people trust me," he says. "Second, I have to sell my company. And, third, I have to sell the product itself." There's no shortcut, I realize, even in a land of 25 million open defecators. It's just toilet by toilet by toilet.

Epilogue: It Hits the Fan

As I finished writing this book, the emerging coronavirus crisis brought many of its themes to the fore and made me consider them in a new way. To my amazement, much of the early reporting during the pandemic centered on toilet paper, as store shelves emptied out and remained vacant for weeks on end. Selfish people were hoarding, many claimed, or just wanted to feel a sense of control in an uncertain time. Others pointed out, rightly, that people simply needed lots more toilet paper than usual if they were going to be home all the time, instead of going to school and to work, and that it wasn't easy for the supply chains to reroute it from businesses (where the rolls are bigger, scratchier, and come in larger quantities) to homes (where the rolls are smaller and plusher). Certainly, people weren't particularly good at estimating *how much* toilet paper they would need. In response, some journalists wrote articles about the wastefulness of toilet paper, pointing to alternatives. Bidet companies reported mass orders; I personally know at least one person who placed one.

Just back from a conference on fatbergs, I worried about sewer health. Without ready access to toilet paper, people were buying more wipes than ever before. Some people flushed them, along with the wipes that they used to clean surfaces and hands. They flushed gloves and masks. This grew to emergency proportions. Philadelphia saw twelve times more clogging than usual at its processing facilities, going from a hundred pounds of waste per year to a hundred pounds per *month*. Wastewater workers fielded late-night calls to unclog the pumps that keep the sewage flowing to the treatment plant. A friend's New York City building had to call a plumber on Easter for a blockage that caused sewage to back up into the basement.

The essential nature of sanitation put workers at risk in a whole new way. Although scientists quickly established that sewage is not a major vector for the spread of the coronavirus, it carries with it lots of other pathogens. Because of the pandemic, the personal protective equipment—masks and gloves—that wastewater workers should use were in short supply. Many in the industry expressed reluctance to prioritize their safety over that of front-line medical workers, but putting their own health at risk didn't seem like an acceptable alternative. In countries with fewer resources, protective equipment was scarcer still. In India, sanitation workers petitioned the government for some simple equipment so that they could keep doing their critical jobs while others were confined to their homes.

For years, people have been warning that the aging of the U.S. professional wastewater workforce will bring a wave of retirements known as the "silver tsunami." But the virus's particular viciousness toward those over sixty brought that into even starker relief. A few wastewater utilities, afraid of losing these critical, hard-to-replace workers to illness or worse, locked down their facilities with the workers inside the gates for weeklong shifts, so that the virus couldn't enter. Others considered how to switch to remote monitoring and other digital approaches in order to limit how often their workers had to go out into the field, where they could come into contact with sewage and potentially infected people. At the same time, already-strained budgets neared a breaking point, with tax revenues declining, business customers disappearing, and record numbers of unemployed customers unable to afford their bills.

Wastewater professionals also learned that they might hold a key to controlling this pandemic and future disease outbreaks. Sewage surveillance might not have gotten much attention had government organizations, public health departments, diagnostic laboratories, and drug companies managed to design, manufacture, and administer a sufficient number of individual tests in time—but, to our great detriment, they did not. (This was often called the original sin of the pandemic response, the one that allowed it to balloon out of control in the United States and some other countries.) So scientists, including Biobot Analytics and Arizona State University's Rolf Halden among many others, began sampling and analyzing sewage arriving at treatment plants, accelerating the scientific process to see if they could provide meaningful feedback to policy makers about how their decisions were affecting the rate of infections in an area. Vik Kashyap also deliberated about how to put his TrueLoo smart toilet to work for the cause, especially in hard-hit senior-care homes.

In Haiti SOIL briefly returned to emergency-response mode, working with the government to place toilet stalls in marketplaces; when the lockdown lifted, they found customers eager to sign up, newly attuned to the importance of private toilets, although the economic effects of the pandemic left them less able to pay. In the United States and Europe, lockdowns forced many public toilets to close, leading to reports of open defecation in public parks and on beaches. And people wondered what would happen when lockdowns lifted. A museum or zoo or workplace or even a restaurant could create sufficient distance between people, but what about when visitors felt the call of nature? As a headline in *Slate* put it, "The Road to Reopening America Runs through the Bathroom."

All of this underscored the importance of toilets—and the centrality of them. The crisis reinforced my respect for the wastewater workforce—and infrastructure—that kept the toilets flushing without interruption. But it also provoked a new sense of urgency about the need for change. The world could see many more pandemics if we don't change how we behave as a species in so many ways, from improving health surveillance to addressing the dangerous environments, such as sewage-laden waterways, where new pathogens could originate. And other crises are looming, many of which touch on sanitation—as a cause, a solution, or both: Plastic pollution. The end of cheap phosphorus. The degradation of agricultural soils. Droughts and storms made worse by climate change. Rising inequality. We have a lot of problems to solve, and there's no time at all to waste, or we'll find ourselves in a dystopian future much faster than we ever imagined.

During the coronavirus crisis, politicians summoned unprecedented sums in short time spans to prop up the world's economies. If they can do that, they can ensure everyone in the world has a safe toilet and make all of our sanitation systems more sustainable—and these investments would pay for themselves, and more, over time. While there are many technological and social challenges ahead, they're far from insurmountable. Compared with many of the world's problems, universal, sustainable sanitation is something that's utterly achievable if it's made a priority. And it's the right thing to do.

As the pandemic began to pick up steam, Indian author and activist Arundhati Roy penned a moving essay that called humanity to see the pandemic as an opportunity to build a better world. "Nothing could be worse than a return to normality," she writes:

> Historically, pandemics have forced humans to break with the past and imagine their world anew. This one is no different. It is a portal, a gateway between one world and the next.
>
>   We can choose to walk through it, dragging the carcasses of our prejudice and hatred, our avarice, our data banks and dead ideas, our dead rivers and smoky skies behind us. Or we can walk through lightly, with little luggage, ready to imagine another world. And ready to fight for it.

What could embody Roy's "normality"—the "good enough" laced with inequality, pollution, and waste—more than the toilet? The device we've inherited arose on the other side of a previous pandemic, that of cholera, which terrorized cities for decades. So it makes sense that our current era of crises, from pandemics to climate change to wealth disparities, could again result in a new kind of toilet, a new kind of *sanitation*, which takes into account not only germs but also the food we eat, the water we drink, the energy we use, the stuff we buy and discard, the nature all around us, and every single one of the billions of people with whom we share this precious planet. People may never march in the streets for better sanitation, but we can hope for a, ahem, quieter movement. It begins, simply enough, by talking more about the toilet—in depth, with sophistication and subtlety, with curiosity and compassion, with urgency and hope, with humor but without embarrassment or shame. The toilet is fascinating, it's ubiquitous, and there is no good future without it.

Notes, Resources, and Further Reading for the Very Curious

Because readers are likely to be more familiar with the U.S. customary system (foot, pound, gallon, et cetera), I use these units throughout, except in a few indicated places. *Ton* (a particularly confusing term) refers in this text to the U.S. ton, or short ton, which equals two thousand pounds.

You can find more resources and an extended reading list at www.chelseawald.com.

Preface

For the amount of pee and poop a person creates, see C. Rose et al., "The Characterization of Feces and Urine: A Review of the Literature to Inform Advanced Treatment Technology," *Crit Rev Environ Sci Technol* 45, no. 17 (2015): 1827–1879. I heard Emily Woods's story on the *Finding Impact Podcast* (#85). Regarding the veracity of the *Hidden Figures* scene, see Dexter Thomas's article "Oscar-Nominated *Hidden Figures* Was Whitewashed—but It Didn't Have to Be," *Vice*, January 25, 2017. I found a 2016 inventory of the bathrooms on Columbia's campus on Bwog.com. Rose George's *The Big Necessity: The Unmentionable World of Human Waste and Why It Matters* (New York: Metropolitan Books, 2009) is essential reading on the topic. The concept of the different ways to use *shit* (if not the specific examples) comes from Nick Haslam's chapter on scatological swearing in *Psychology in the Bathroom* (Houndmills, Basingstoke, Hampshire, UK: Palgrave Macmillan, 2012). *Fast Company* published a thorough treatment of the poop emoji called "The Oral History of the Poop Emoji (or, How Google Brought Poop to America)," by Lauren Schwartzberg, November 18, 2014.

The New Toilet Revolution

Chapter epigraph from John Harington, *A New Discourse of a Stale Subject, Called the Metamorphosis of Ajax*, ed. Elizabeth Story Donno (New York: Columbia University Press, 1962), quoted in Dolly Jørgensen, "The Metamorphosis of Ajax, Jakes, and Early Modern Urban Sanitation," *Early English Studies* 3 (2010): 1–29.

Getting to Loo-topia: More on the *British Medical Journal*'s rankings can be read in its pages in the article "*BMJ* Readers Choose the 'Sanitary Revolution' as Greatest Medical Advance since 1840" from 334, no. 7585 (January 20, 2007): 111. The Buckminster Fuller quote comes from his book *I Seem to Be a Verb: Environment and Man's Future* (New York: Bantam Books, 1970). The eye-opening 2019 report on the state of water and sanitation in the United States, from the U.S. Water Alliance, along with DigDeep and Michigan State University, is called *Closing the Water Access Gap in the United States: A National Action Plan*; Catherine Coleman Flowers's book is *Waste: One Woman's Fight Against America's Dirty Secret* (New York: New Press, 2020). See also Inga Winkler, Catherine Coleman Flowers, and JoAnn Kamuf Ward, *Flushed and Forgotten: Sanitation and Wastewater in Rural Communities in the United States* (Alabama Center for Rural Enterprise, Columbia Law School Human Rights Clinic, Institute for the Study of Human Rights at Columbia University, 2019) and Drew Capone et al., "Water and Sanitation in Urban America, 2017–2019," *American Journal of Public Health* 110, no. 10 (2020): 1567–1572. Research on the Black Belt is detailed in Megan L. McKenna et al., "Human Intestinal Parasite Burden and Poor Sanitation in Rural Alabama," *American Journal of Tropical Medicine and Hygiene* 97, no. 5 (2017): 1623–1628; I also recommend the coverage of this issue in the *Guardian* and *Southerly*. I first learned about the cesspools in Hawaii from "Finally Tackling a Crappy Situation," by Stuart Coleman, in *Hawaii Business*, September 5, 2019. The Dutch expenditure on water and sanitation, relative to others, is in *Financing Water Supply, Sanitation, and Flood Protection: Challenges in EU Member States and Policy Options* (Paris: OECD Publishing, 2020). References for the Sneek example include Tove A. Larsen, Kai M. Udert, and Judit Lienert, eds., *Source Separation and Decentralization for Wastewater Management* (London and New York: IWA Publishing, 2013), and Dries Hegger and Bas J. M. Van Vliet, "End User Perspectives on the Transformation of Sanitary Systems," in *Social Perspectives on the Sanitation Challenge*, eds. Bas van Vliet, Gert Spaargaren, and Peter Oosterveer (Dordrecht: Springer, 2010), 203–216.

In general, I use and recommend the Sustainable Sanitation and Water Management Toolbox (sswm.info) as a reference for learning more about just about any type of toilet system.

What's in the Toilet?: The main source for this section is the previously mentioned Rose et al., "The Characterization of Feces and Urine." For the microbes in poop, I also referred to sources including Vincent Ho's article in *The Conversation*, "Your Poo Is (Mostly) Alive. Here's What's in It," October 31, 2018; Robert L. Atmer et al., "Norwalk Virus Shedding after Experimental Human Infection," *Emerging Infectious Diseases* 14, no. 10 (2008): 1553–1557; the World Bank report by Richard G. Feachum et al., *Sanitation and Disease: Health Aspects of Excreta and Wastewater Management* (Chichester, UK: John Wiley & Sons, 1983); and Elizabeth Thursby and Nathalie Juge, "Introduction to the Human Gut Microbiota," *Biochem J.* 474, no. 11 (2017): 1823–1836.

A Very Short History of Toilet Systems: For this history (and mentions of history in the rest of the book), I drew on many sources, especially these books: James L. A. Webb, Jr., *The Guts of the Matter: A Global History of Human Waste and Infectious Intestinal Disease* (Cambridge: Cambridge University Press, 2020); Martin V. Melosi, *The Sanitary City: Environmental Services in Urban America from Colonial Times to the Present* (Baltimore: Johns Hopkins University Press, 2008); Stephen Halliday, *The Great Stink of London: Sir Joseph Bazalgette and the Cleansing of the Victorian Metropolis* (Stroud, UK: Sutton, 1999); David Eveleigh, *Privies and Water Closets* (Oxford: Shire, 2008); and Ann Olga Koloski-Ostrow, *The Archaeology of Sanitation in Roman Italy: Toilets, Sewers, and Water Systems* (Chapel Hill: University of North Carolina Press, 2015). In addition, I referred to Andreas N. Angelakis and Joan B. Rose, eds., *Evolution of Sanitation and Wastewater Technologies through the Centuries* (London: IWA Publishing, 2014); Piers D. Mitchell, ed., *Sanitation, Latrines and Intestinal Parasites in Past Populations* (London: Taylor and Francis, 2016); and Jørgensen, "The Metamorphosis of Ajax, Jakes, and Early Modern Urban Sanitation." Some of the reporting comes originally from my May 24, 2016, *Nature* feature, "The Secret History of Ancient Toilets," edited with care by Rich Monastersky.

For global sanitation data and reports describing the various levels of sanitation, refer to washdata.org; a good overview specifically for cities is D. Satterthwaite, V. A. Beard, D. Mitlin, and J. Du, "Untreated and Unsafe: Solving the Urban Sanitation Crisis in the Global South," Working Paper (Washington, D.C.: World Resources Institute, 2019). For primers on fecal sludge and what to do with it, see Linda Strande's online course from Coursera, Introduction to Faecal Sludge Management, and Strande, Mariska Ronteltap, and Damir Brdjanovic, eds., *Faecal Sludge Management: Systems Approach for Implementation and Operation* (London: IWA Publishing, 2014). "Sludge bomb" comes from Tilley's presentation at the 2019 Faecal Sludge Management conference in Cape Town, South Africa. Narain quote from "The Flush Toilet Is Ecologically Mindless," *Down to Earth* 10, no. 19 (2002). The 2019 World Bank report mentioned is Julie Rozenberg and Marianne Fay, *Beyond the*

Gap: How Countries Can Afford the Infrastructure They Need while Protecting the Planet, Sustainable Infrastructure series (Washington, D.C.: World Bank, 2019); see also Amanda Goksu et al., *Reform and Finance for the Urban Water Supply and Sanitation Sector* (Washington, D.C.: World Bank, 2019); and Luis A. Andres et al., *Doing More with Less: Smarter Subsidies for Water Supply and Sanitation* (Washington, D.C.: World Bank, 2019). For return on investment in sanitation, Guy Hutton, "Global Costs and Benefits of Reaching Universal Coverage of Sanitation and Drinking-Water Supply," *J Water Health* 11, Issue 1 (2013): 1–12. Quote about declining federal funding from *Closing the Water Access Gap*. Climate-related figures and further reading from F. Mills et al., *Considering Climate Change in Urban Sanitation: Conceptual Approaches and Practical Implications* (USHHD Learning Brief) (The Hague: SNV, 2019), and the SuSanA background paper "Opportunities for Sustainable Sanitation in Climate Action" (2019). Parker Thomas Moon quote from Barbara Penner's *Bathroom (Objekt)* (London: Reaktion Books, 2013).

The New Toilet Science: My source on Maximum Performance is Ariel Schwartz's "Designing the Toilet of the Future? You're Going to Need This Fake Poop," *Fast Company*, November 19, 2013; the Eawag synthetic poop paper is Roni Penn et al., "Review of Synthetic Human Faeces and Faecal Sludge for Sanitation and Wastewater Research," *Water Research* 132 (2017): 222–240. Background and more on shit flow diagrams at sfd.susana.org; Barbara Evans's quote from TEDxMacclesfield talk "Cutting the Crap: Saving Lives by Solving the Right Problem" (2019); public toilet rented as church in Anthony Langat's "When Public Restrooms Fail, Rent Them Out as Churches?," *Devex*, February 19, 2019.

The Transformed Toilet: More on the gap between idea and large-scale implementation in Naomi Carrard and Juliet Willets, "Environmentally Sustainable WASH? Current Discourse, Planetary Boundaries and Future Directions," *Journal of Water, Sanitation and Hygiene for*

Development 7, no. 2 (2017): 209–228, and Anne-Katrin Skambraks et al., "Source Separation Sewage Systems as a Trend in Urban Wastewater Management: Drivers for the Implementation of Pilot Areas in Northern Europe," *Sustainable Cities and Society* 28 (2017).

Paging Dr. Toilet

Chapter epigraph from John Gregory Dunne, *Nothing Lost* (New York: Knopf, 2004).

It Takes Guts: For more on Vik Kashyap's medical story before Toi Labs, see Moises Velasquez-Manoff, *An Epidemic of Absence: A New Way of Understanding Allergies and Autoimmune Diseases* (New York: Scribner, 2012), in which Kashyap is called by the pseudonym Rick; the scientific article that Kashyap co-authored is Mara J. Broadhurst et al., "IL-22+ CD4+ T Cells Are Associated with Therapeutic *Trichuris Trichiura* Infection in an Ulcerative Colitis Patient," *Science Translational Medicine* 2, no. 60 (2010): 60ra88.

For facts about the pathogens and diseases mentioned in this chapter, I relied on the CDC's and WHO's excellent fact sheets. For background knowledge about sanitation-related diseases, I am indebted to Tufts University professors Jeffrey Griffiths and David M. Gute, whose online course, Biology of Water and Health, I took on the EdX platform. For history in this section, I referred to *The Guts of the Matter*, *The Great Stink of London*, *The Sanitary City*, *Privies and Water Closets*, and *Sanitation, Latrines and Intestinal Parasites in Past Populations*, as well as Steven Johnson's *The Ghost Map: The Story of London's Most Terrifying Epidemic—and How It Changed Science, Cities, and the Modern World* (New York: Riverhead Books, 2006). For Valerie Curtis on disgust, I referred to *Don't Look, Don't Touch, Don't Eat: The Science Behind Revulsion* (Chicago: University of Chicago Press, 2013); "Dirt, Disgust and Disease: A Natural History of Hygiene," *J Epidemiol Community Health* 61, no. 8 (2007): 660–664;

and Curtis, Mícheál de Barra, and Robert Aunger, "Disgust as an Adaptive System for Disease Avoidance Behaviour," *Philosophical Transactions of the Royal Society B* 366 (2011): 389–401. Bible scholarship from Jodi Magness, "What's the Poop on Ancient Toilets and Toilet Habits?," *Near Eastern Archaeology* 75, no. 2 (2012): 80–87.

Lost in Translation: Enders's book is *Gut* (Vancouver: Greystone Books, 2015). Kira's influential study, now unfortunately out of print, was first published as a report by Cornell University in 1966 and then in paperback as *The Bathroom: Criteria for Design* (New York: Bantam Books, 1967). The Ohio State study is Rohan M. Modi et al., "Implementation of a Defecation Posture Modification Device: Impact on Bowel Movement Patterns in Healthy Subjects," *Journal of Clinical Gastroenterology* 53, no. 3 (2019): 216–219; see also Klaus Goeschen and Bernhard Liedl, "Chronic Pelvic Pain and Pelvic Organ Prolapse: A Consequence of Upright Position?," *Pelviperineology* 39 (2020): 55–59; Pankaj Garg, "Physiologic Management of Chronic Constipation: Let's FEED It," *Dig Dis Sci* 62 (2017): 3254–3255; and S. Takano and D. R. Sands, "Influence of Body Posture on Defecation: A Prospective Study of 'The Thinker' Position," *Tech Coloproctol* 20 (2016): 117–121. The restroom bacteria study is S. M. Gibbons et al., "Ecological Succession and Viability of Human-Associated Microbiota on Restroom Surfaces," *Appl Environ Microbiol* 81 (2015): 765–773. On bidets, see Shannon Palus, "Are Bidets Better for You than Toilet Paper?," The Wirecutter, September 7, 2016; *Berkeley Wellness Letter*'s "Ask the Experts: Bidets," January 1, 2013; K. Asakura et al., "Relationship between Bidet Toilet Use and Haemorrhoids and Urogenital Infections: A 3-Year Follow-Up Web Survey," *Epidemiology and Infection* 146, no. 6 (2018): 763–770; Mitsuharu Ogino, "Habitual Use of Warm-Water Cleaning Toilets Is Related to the Aggravation of Vaginal Microflora," *Journal of Obstetrics and Gynaecology Research* 36, no. 5 (2010): 1071–1074; Akira Tsunoda et al., "Survey of Electric Bidet Toilet Use among Community Dwelling Japanese People and Correlates for an Itch on the Anus," *Environ Health Prev Med* 21

(2016): 547–553; and A. Kanayama Katsuse, "Public Health and Healthcare-Associated Risk of Electric, Warm-Water Bidet Toilets," *Journal of Hospital Infection* 97, no. 3 (2017): 296–300.

The German toilet manufacturer referred to is Reuter. Poop inspection studies: K. W. Heaton et al., "Defecation Frequency and Timing, and Stool Form in the General Population: A Prospective Study," *Gut* 33 (1992): 818–824, and M. R. Blake et al., "Validity and Reliability of the Bristol Stool Form Scale in Healthy Adults and Patients with Diarrhoea-Predominant Irritable Bowel Syndrome," *Alimentary Pharmacology and Therapeutics* 44, no. 7 (2016): 693–703.

A Crying Shame: Kamal Kar has told his story many times in public; for this, and in particular for the quotes, I draw largely on his appearance in the Coursera course titled Water Supply and Sanitation Policy in Developing Countries Part 2: Developing Effective Interventions, offered by the University of Manchester and taught by Duncan Thomas and Dale Whittington. The training manual is Kamal Kar with Robert Chambers, *Handbook on Community-Led Total Sanitation* (Brighton: IDS, 2008).

In addition to CDC and WHO sources, for current data on illness and death, as well as some related WASH statistics, I turned to Annette Prüss-Ustün et al., "Burden of Disease from Inadequate Water, Sanitation and Hygiene for Selected Adverse Health Outcomes: An Updated Analysis with a Focus on Low- and Middle-Income Countries," *International Journal of Hygiene and Environmental Health* 222, no. 5 (2019): 765–777. On the connection between poor sanitation and stunting, see, among others, Sophie Budge et al., "Environmental Enteric Dysfunction and Child Stunting," *Nutrition Reviews* 77, no. 4 (2019): 240–253, and Dean Spears, "Exposure to Open Defecation Can Account for the Indian Enigma of Child Height," *Journal of Development Economics* 146 (2020). For technical insights on antibiotic resistance, see WHO, FAO, and World Organization for Animal Health, *Technical Brief on Water, Sanitation, Hygiene and*

Wastewater Management to Prevent Infections and Reduce the Spread of Antimicrobial Resistance (Geneva: WHO, 2020). On the study of rivers, which was presented at a 2019 conference, see, for example, Natasha Gilbert's "World's Rivers 'Awash with Dangerous Levels of Antibiotics'" in the *Guardian* (May 26, 2019).

Alexandra Brewis and Amber Wutich's book, *Lazy, Crazy, and Disgusting: Stigma and the Undoing of Global Health* (Baltimore: Johns Hopkins University Press, 2019), summarizes the case against CLTS and points to many additional sources. For the disappointing WASH intervention trials and resulting discussion, see, for example, Oliver Cumming et al., "The Implications of Three Major New Trials for the Effect of Water, Sanitation and Hygiene on Childhood Diarrhea and Stunting: A Consensus Statement," *BMC Medicine* 17, art. no. 173 (2019); Amy J. Pickering et al., "The WASH Benefits and SHINE Trials: Interpretation of WASH Intervention Effects on Linear Growth and Diarrhoea," *The Lancet Global Health* 7 (2019): E1139–E1146; and (for the quote referring to the FDA) Joe Brown, Jeff Albert, and Dale Whittington, "Community-Led Total Sanitation Moves the Needle on Ending Open Defecation in Zambia," *American Journal of Tropical Medicine and Hygiene* 100, no. 4 (2019): 767–769.

Test the Waters: In writing the story of Biobot Analytics, I referred to sources including Norkio Endo et al., "Rapid Assessment of Opioid Exposure and Treatment in Cities through Robotic Collection and Chemical Analysis of Wastewater," *Journal of Medical Toxicology* 16 (2020): 195–203; Claire Duvallet et al., "Mapping Community Opioid Exposure through Wastewater-Based Epidemiology as a Means to Engage Pharmacies in Harm Reduction Efforts," *Preventing Chronic Disease* 17, no. 200053 (2020); Donald Smith, "Building Smarter Healthier Communities through Opioid Wastewater Monitoring," *NC Currents*, Fall 2019, 81–84; Celia Henry Arnaud, "Mariana Matus Means to Combat the Opioid Epidemic with Chemical Data," *Chemical & Engineering News* 98, no. 9 (2020); Kathy Pretz, "Combating the

Opioid Crisis, One Flush at a Time," *IEEE Spectrum*, July 3, 2019; and Justin Chen, "Scientists Can Track the Spread of Opioids in Sewers. But Do Cities Want to Know What Lies Below?," *STAT*, June 26, 2018. The company's "public call to action" on Medium.com is "Testing for COVID-19 beyond the Clinic: Using Wastewater Epidemiology to Proactively Detect Outbreaks," March 12, 2020. Another of Arnaud's *C&EN* stories, "To Monitor the Health of Cities' Residents, Look No Further than Their Sewers," April 30, 2018, also proved useful to understanding the backstory of wastewater epidemiology, while, for further technical information, I turned to the European Monitoring Centre for Drugs and Drug Addiction's book *Assessing Illicit Drugs in Wastewater: Advances in Wastewater-Based Drug Epidemiology* (Lisbon: EMCDDA, 2016). The Italian study mentioned is Ettore Zuccato et al., "Cocaine in Surface Waters: A New Evidence-Based Tool to Monitor Community Drug Abuse," *Environmental Health* 4, no. 14 (2005). Thanks additionally to Frederic Béen at KWR in the Netherlands for taking the time to explain KWR's wastewater surveillance work to me and show me the laboratory.

For the promise of urine metabolites (as well as the prototype from the University of Wisconsin–Madison in the next section), see Ian J. Miller et al., "Real-Time Health Monitoring through Urine Metabolomics," *npj Digital Medicine* 2 (2019): 109. Rolf Halden gave a lecture titled "Urban Metabolism Metrology: A New Discipline Elucidating the Human Condition in Cities around the World" at the opening session of the Fall 2016 American Chemical Society National Meeting (available online). His related publications include "On the Need and Speed of Regulating Triclosan and Triclocarban in the United States," *Environmental Science & Technology* 48, no. 7 (2014): 3603–3611; Arjun Venkatesan and Halden, "Wastewater Treatment Plants as Chemical Observatories to Forecast Ecological and Human Health Risks of Manmade Chemicals," *Scientific Reports* 4, art. no. 3731 (2015); Venkatesan et al., "United States National Sewage Sludge Repository at Arizona State University: A New Resource and Research Tool for

Environmental Scientists, Engineers, and Epidemiologists," *Environ Sci Pollut Res Int* 22, no. 3 (2015): 1577–1586; and Olga E. Hart and Halden, "Computational Analysis of SARS-CoV-2/COVID-19 Surveillance by Wastewater-Based Epidemiology Locally and Globally: Feasibility, Economy, Opportunities and Challenges," *Science of the Total Environment* 730, art. no. 138875 (2020). In 2020, Halden gave a detailed interview to Craig Gustavson that appears in *Integrative Medicine* 19, no. 1, 52–56, and published a personal, wide-ranging book titled *Environment (Object Lessons)* (New York: Bloomsbury).

The role of sewage in polio eradication in Marisa Eisenberg, Andrew Brouwer, and Joseph Eisenberg, "Sewage Surveillance Is the Next Frontier in the Fight against Polio," *The Conversation*, October 19, 2018; Jacob Moran-Gilad et al., "Field Study of Fecal Excretion as a Decision Support Tool in Response to Silent Reintroduction of Wild-Type *Poliovirus* 1 into Israel," *Journal of Clinical Virology* 66 (2015): 51–55; Andrew F. Brouwer et al., "Epidemiology of the Silent Polio Outbreak in Rahat, Israel, Based on Modeling of Environmental Surveillance Data," *Proceedings of the National Academy of Sciences* 115, no. 45 (2018): E10625–E10633; and polioeradication.org.

Data Stream: Some reporting in this chapter (and especially this section) originated with my *Nature Outlook* feature, "Diagnostics: A Flow of Information," November 9, 2017; special thanks to Michelle Grayson for her helpful editing. I learned about the VIP toilet from "Stay Healthy with the Toilet 'Doctor,'" *BBC News*, July 11, 2001. The Google patent is U.S. No. 10,064,582. For Gambhir's quest, "'And Yet, You Try'" by Julie Gericius in *Stanford Medicine*, Fall 2016; his own opinion essay in *STAT*, titled "New Ways to Detect Cancer Early Will Help Pave the Way for Precision Health," March 4, 2019; Gambhir et al., "Toward Achieving Precision Health," *Science Translational Medicine* 10, no. 430 (2018): eaao3612; and Seung-min Park, "Precision Health Toilet: More than Saving Toilet Paper," *Nature Research Bioengineering Community* (blog), April 6, 2020. Risks of wearables in Lukasz Piwek et al., "The Rise of Consumer Health Wearables: Promises and Barriers," *PLoS Med* 13, no. 2 (2016): e1001953. The *Washington Post* article in which the Kim Jong-un quote seems to have originated is Anna Fifield's "South Korea Is Sparing No Effort to Make Summit with Kim a Made-for-TV Success," April 25, 2018.

Pipe Down

Chapter epigraph from Victor Hugo, *Les Misérables*, trans. Isabel Florence Hapgood (New York: Thomas Y. Crowell & Co., 1887; Urbana, Illinois: Project Gutenberg, 2008).

System Critical: Roman references in this and the next section in Koloski-Ostrow's book. Parts of this chapter were adapted from my February 28, 2019, feature "Cities without Sewers" for *Rethink*, an online publication of the Stockholm Resilience Centre. The feature was one part of a European Journalism Centre–funded project called Waste Not. (See end for details about the grant.) Thank you to Naomi Lubick for superb editing on that story; Daniel Tillias, who translated for me in Haiti; Caleb Alcénat of Labelimage, who took photographs; and the entire team at SOIL.

For the history of sewers, I turned particularly to *The Sanitary City* and *The Great Stink of London*, as well as to Christopher Hamlin, "Edwin Chadwick and the Engineers, 1842–1854: Systems and Antisystems in the Pipe-and-Brick Sewers War," *Technology and Culture* 33, no. 4 (1992): 680–709.

Redesigning from the Bottom Up: For some of this section, I drew on my reporting for my November 7, 2013, feature for *Nautilus* magazine, "Lavatory Laboratory," assigned by Kevin Berger and edited by Amy Maxmen (many thanks to both for helping to launch me on this toilet journey), and I also learned a lot from an Eawag course, offered on Coursera, called Planning & Design of Sanitation Systems and Technologies, run by Christoph Lüthi. For an overview of distributed water and waste-water infrastructure, as well as some details and examples mentioned in this section, I turned to the Broadview Collaborative's 2019 report *Opportunities in Distributed Water Infrastructure*. For sobering reads on the world's sewage "white elephants," see WaterAid's 2019 report *Functionality of Wastewater Treatment Plants in Low- and Middle-Income Countries: Desk Review* and the Eawag-led M. Klinger, A. Gueye, A. Manandhar Sherpa, and L. Strande, *Scoping Study: Faecal Sludge Treatment Plants in South-Asia and Sub-Saharan Africa,* eFSTP Project Report, 2019. For the Gates Foundation and its founders' approach to sanitation, I relied on sources such as Sylvia Mathews Burwell's 2011 speech to an African Ministers' Council on Water conference, Bill Gates's 2018 speech to the Reinvented Toilet Expo, the GatesNotes blog, and the first episode of the 2019 Netflix series *Inside Bill's Brain*. For academic critiques of the Gates Foundation's approach in other sectors, see Rachel Schurman, "Micro(soft) Managing a 'Green Revolution' for Africa: The New Donor Culture and International Agricultural Development," *World Development* 112 (2018): 180–192, and Anne-Emanuelle Birn and Judith Richter, "U.S. Philanthrocapitalism and the Global Health Agenda: The Rockefeller and Gates Foundations, Past and Present," in *Health Care under the Knife: Moving beyond Capitalism for Our Health,* ed. Howard Waitzkin (New York: Monthly Review Press, 2018). On alterna-tive sewer systems, Marius Mohr et al., "Vacuum Sewerage Systems—a Solution for Fast Growing Cities in Developing Countries?," *Water Practice & Technology* 13, no. 1 (2018): 157–163a; Jose Carlos Melo, *The Experience of Condominial Water and Sewerage Systems in Brazil: Case Studies from Brasilia, Salvador and Parauebas* (Lima, Peru: World Bank, BNWP, WSP, 2005); and Mingma Gyalzen Sherpa et al., "Applying the Household Centered Environmental Sanitation Planning Approach: A Case Study from Nepal," *Journal of Water, Sanitation and Hygiene for Development* 2, no. 2 (2012): 124–132.

The Dirtiest Jobs: Example press reports on sanitation-worker deaths: *India Today*'s "Sanitation Worker Chokes on Toxic Fumes

While Cleaning Sewer in Delhi's Shahdara," February 3, 2020, and Damini Nath's "110 Deaths by Cleaning Sewers, Septic Tanks in 2019," *The Hindu*, February 11, 2020. The mentioned report on sanitation workers is the World Bank, the International Labour Organization, WaterAid, and the World Health Organization, *Health, Safety and Dignity of Sanitation Workers: An Initial Assessment* (Washington, D.C.: World Bank, 2019); it was accompanied by a moving online interactive feature at sanitationwork .wateraid.org. One good source on scavenging in India is Diane Coffey and Dean Spears, *Where India Goes: Abandoned Toilets, Stunted Development and the Costs of Caste* (Noida, Uttar Pradesh: HarperLitmus, 2017). You can see Sharada's photo-essays at SharadaPrasad.com. On the success of the Swachh Bharat Mission, see Val Curtis, "Explaining the Outcomes of the 'Clean India' Campaign: Institutional Behaviour and Sanitation Transformation in India," *BMJ Global Health* 4 (2019): e001892. The *Washington Post* op-ed, from September 6, 2019, is called "The Gates Foundation Shouldn't Give an Award to Narendra Modi." References for Wilson include Divya Trivedi's "Clean Your Mind, Not Our Feet," in *Frontline* (2019), and "Safai Karmachari Andolan: An Insider's Account (Conversation with Bezwada Wilson)" in *The Right to Sanitation in India: Critical Perspectives*, eds. Philippe Cullet, Sujith Koonan, and Lovleen Bhullar (New Delhi: Oxford University Press, 2020).

The New Night Soil: Reports and articles on SOIL's work include: David Berendes et al., "Ascaris and Escherichia Coli Inactivation in an Ecological Sanitation System in Port-au-Prince, Haiti," *PLoS One* 10, no. 5 (2015): e0125336; Gavin McNicol et al., "Climate Change Mitigation Potential in Sanitation via Off-Site Composting of Human Waste," *Nature Climate Change* 10 (2020): 545–549; and the World Bank's "Evaluating the Potential of Container-Based Sanitation" (Washington, D.C.: World Bank, 2019), which includes a detailed case study of SOIL. On South Africa, see Steven Robins, "Poo Wars as Matter Out of Place: 'Toilets for Africa' in Cape Town," *Anthropology*

Today 30, no. 1 (2014); Barbara Penner has also written extensively on the topic, including "Flush with Inequality: Sanitation in South Africa" in *Places Journal*, November 2010. Thanks especially to Gavin McNichol, Rebecca Ryals, and Bernelle Verster.

Taking the Piss

Chapter epigraph from Bret Easton Ellis, *American Psycho* (New York: Vintage Books, 1991).

Urine Heaven: Much of the information on Mothers for Mothers in this chapter, as well as some history of urine use, is from Charlotte Kroløkke's *Global Fluids: The Cultural Politics of Reproductive Waste and Value* (New York: Berghahn Books, 2018). More on the history and culture of hCG in reproductive medicine in Bruno Lunenfeld, "Gonadotropin Stimulation: Past, Present, and Future," *Reproductive Medicine and Biology* 11, no. 1 (2012): 11–25; Lunenfeld et al., "The Development of Gonadotropins for Clinical Use in the Treatment of Infertility," *Front Endocrinol* 10 (2019): 429; and Sandra Bärnreuther, "From Urine in India to Ampoules in Europe: The Relational Infrastructure of Human Chorionic Gonadotropin," *Zeitschrift für Ethnologie* 143 (2018): 41–60.

Guidelines and manuals for home pee-cycling of urine into fertilizer include: A. Richert et al., *Practical Guidance on the Use of Urine in Crop Production* (Stockholm: Stockholm Environmental Institute, 2010); Caroline Schönning, *Urine Diversion—Hygienic Risks and Microbial Guidelines for Reuse* (Solna: Swedish Institute for Infectious Disease Control, n.y.); and "Using Urine as a Fertilizer in Home Gardens: Frequently Asked Questions" from the Rich Earth Institute (2019). Academic sources on the historical uses of urine include Andrew Wilson and Miko Flohr's chapter "The Economy of Odure" in *Roman Toilets: Their Archaeology and Cultural History*, eds. Gemma C. M. Jansen, Ann O. Koloski-Ostrow, and Eric C. Moormann (Leuven: Peeters, 2011). The Vespasian quote is from *The Lives of the*

Twelve Caesars by C. Suetonius Tranquillus, translated by J. C. Rolfe and published in the Loeb Classical Library (1914). See also Xiaoli Ji et al., "Urine-Derived Stem Cells: The Present and the Future," *Stem Cells International* (2017): 4378947. The story in *Slate* is "The Yellow Liquid Diet" by Chris Wilson, May 21, 2008.

Drinking from the Bowl: For this section, I referred often to David Sedlak's *Water 4.0: The Past, Present, and Future of the World's Most Vital Resource* (New Haven: Yale University Press, 2014) and also relied on the excellent reporting in *Circle of Blue*, especially by Brett Walton (the influence of these sources is not, however, limited to this section). For predictions on water-stressed regions and the future of water in general, see UNESCO, UN-Water, *United Nations World Water Development Report 2020: Water and Climate Change* (Paris: UNESCO, 2020). Toilet-flushing water estimate from paper on the mentioned Teflon-like coating: Jing Wang et al., "Viscoelastic Solid-Repellent Coatings for Extreme Water Saving and Global Sanitation," *Nature Sustainability* 2 (2019): 1097–1105. Aside from what I read in media coverage (including News24's January 28, 2018, story "South Africa: Capetonians Might Have to Do 'Royal Flush' Soon"), I learned about the Cape Town drought's effect on wastewater treatment plants on a tour offered by the FSM5/AfricaSan conference in 2019.

Sources on wastewater use and reuse include Manzoor Qadir et al., "Global and Regional Potential of Wastewater as a Water, Nutrient and Energy Source," *Natural Resources Forum* (2020): 1–12; A. L. Thebo et al., "A Global, Spatially-Explicit Assessment of Irrigated Croplands Influenced by Urban Wastewater Flows," *Environmental Research Letters* 12, no. 7 (2017): 074008; Seth M. Siegel's *Let There Be Water: Israel's Solution for a Water-Starved World* (New York: Thomas Dunne Books, 2015); Jacques Leslie's "Where Water Is Scarce, Communities Turn to Reusing Wastewater" in *Yale Environment 360*, May 1, 2018; and (for the quote) ABC7's "OC's 'Toilet to Tap' Drinking Water a Tough Sell Even on a Hot Day," June 21, 2017. Thanks to Niels Groot, of HZ University of Applied Sciences in the Netherlands and Dow Benelux, for explaining the Terneuzen project.

Mayor Pete's Smart Sewers: Se Branko Kerkez, "Smarter Stormwater Systems," *Environmental Science & Technology* 50 (2016): 7267–7273, and Luis Montestruque and M. D. Lemmon, "Globally Coordinated Distributed Storm Water Management System," in *CySWater'15: Proceedings of the 1st ACM International Workshop on Cyber-Physical Systems for Smart Water Networks* (New York: ACM Press, 2015): 1–6. The *South Bend Tribune*'s coverage of the city's sewers provided valuable information for this section, as did John Nagy's "A Sewer System to Make Others Flush with Envy," *Notre Dame Magazine*, Winter 2010–11, and the H2duo's podcast episode #47, "Service, Smart Sewers & Slaying the Basics," March 18, 2019, with Buttigieg as a guest. Buttigieg's book is *Shortest Way* (New York: Liveright, 2019). The "Poop Nightmare" story, a worthy read by Emily Atkin, appeared in the *New Republic* on September 14, 2017. The *Rolling Stone* magazine article is Bob Moser's "Could This 36-Year-Old Indiana Mayor Topple Trumpism?," August 23, 2018.

Yes Wee Can: Massimino quote from the video "STS-132 Atlantis behind the Scenes 1: Space Toilet," May 7, 2010, where you can see the training toilet at work. I first heard about the Erdos case in the online course Planning & Design of Sanitation Systems and Technologies (mentioned earlier); I also referred to the case study *Urine Diversion Dry Toilets in Multi-Storey Buildings*, published by SuSanA in 2012. Waterless urinal background from Kathleen Kokosinski, "Entrepreneurial Biography: Klaus Reichardt, Founder/Managing Partner of Waterless Co. LLC," in *Frontiers in Eco-Entrepreneurship Research*, vol. 20, ed. Gary D. Libecap (Bingley, UK: Emerald Group Publishing, 2009): 1–14, and Joshua Davis, "Pissing Match: Is the World Ready for the Waterless Urinal?," *Wired*, June 22, 2010.

On women and peeing/urinals/Lapee at festivals, Malene Enø's "We Asked a Bunch of Girls about Peeing, Quite Possibly the Greatest Feat

You Can Accomplish at Roskilde Festival," in *Girls Are Awesome*, June 30, 2017; Elizabeth Segran's "The Quest to Bring Urinal Culture to Women," *Fast Company*, July 19, 2019; and Richard Orange's "The Penny Drops: At Last a Female Urinal for the Festival Crowd," in the *Guardian*, July 4, 2019 (also the source of the quote). Olmert's book *Bathroom's Make Me Nervous: A Guidebook for Women with Urination Anxiety (Shy Bladder)* (Walnut Creek: CJOB, 2008). Case wrote the essay "Why Not Abolish the Laws of Urinary Segregation?" in the invaluable *Toilet: Public Restrooms and the Politics of Sharing*, eds. Harvey Molotch and Laura Norén (New York: NYU Press, 2010).

I first interviewed Gründl for and wrote about the Blue Diversion toilet in "Lavatory Laboratory." His toilet projects' publications include Joel Gundlach et al., "Novel NoMix Toilet Concept for Efficient Separation of Urine and Feces and Its Design Optimization Using Computational Fluid Mechanics," *Journal of Building Engineering*, art. no. 101500 (2020); Robert Tobias et al., "Early Testing of New Sanitation Technology for Urban Slums: The Case of the Blue Diversion Toilet," *Science of the Total Environment* 576 (2017): 264–272; and Tove A. Larsen et al., "Blue Diversion: A New Approach to Sanitation in Informal Settlements," *Journal of Water, Sanitation and Hygiene for Development* 5, no. 1 (2014): 64–71.

Pee Is for Precious: "Black box" paper is Claudiu-Eduard Nedelciu et al. "Opening Access to the Black Box: The Need for Reporting on the Global Phosphorus Supply Chain," *Ambio* 49 (2020): 881–891; other useful sources on phosphorus include J. J. Mortvedt, "Heavy Metal Contaminants in Inorganic and Organic Fertilizers," *Fertilizer Research* 43 (1995): 55–61, and Mesfin M. Mekonnen and Arjen Y. Hoekstra, "Global Anthropogenic Phosphorus Loads to Freshwater and Associated Grey Water Footprints and Water Pollution Levels: A High-Resolution Global Study," *Water Resources Research* 54, no. 1 (2017): 245–358. For details on the Rich Earth Institute's program, Jennifer Atlee et al., *Guide to Starting a Community-scale Urine*

Diversion Program (Brattleboro, VT: Rich Earth Institute, 2019). For urine-powered mobile phones, see Ieropoulos et al., "Waste to Real Energy: The First MFC Powered Mobile Phone," *Physical Chemistry Chemical Physics* 15 (2013): 15312–15316, as well as later publications.

For the history of pregnancy tests, the National Institutes of Health has an informative online exhibit called *The Thin Blue Line: The History of the Pregnancy Test*; regarding the role in chytrid, Vance T. Vredenburg et al., "Prevalence of Batrachochytrium dendrobatidis in Xenopus Collected in Africa (1871–2000) and in California (2001–2010)," *PLoS ONE* 8, no. 5 (2013): E63791, and Michael Greshko, "Amphibian 'Apocalypse' Caused by Most Destructive Pathogen Ever," *National Geographic*, March 28, 2019.

Eating Sh!t

Chapter epigraph found in Haslam's *Psychology in the Bathroom* and traced through K. Reiskel, "Skatologische Inschriften," *Anthropophyteia* 3 (1906): 244–246, to *Kryptadia: Recueil de document pour servir á l'étude des traditions populaires VI* (Paris: H. Welter, 1899). Versions of this verse remain in circulation in France today.

Black Gold: There are lots of good sources on the Boston Harbor Cleanup, including the pelletizing aspect; I found the following especially useful: Paul S. Levy and Michael S. Connor, "The Boston Harbor Cleanup," *New England Journal of Public Policy* 8, no. 2, art. no. 7 (1992); Di Jin et al., "Evaluating Boston Harbor Cleanup: An Ecosystem Valuation Approach," *Frontiers in Marine Science* 5 (2018): 478; and James DeCoq, Kirsty Gray, and Robert Churchill, "Sewage Sludge Pelletization in Boston: Moving Up the Pollution Prevention Hierarchy," October 1998. General sources on sludge reuse include Roland J. LeBlanc, Peter Matthews, and Roland P. Richard, eds., *Global Atlas of Excreta, Wastewater Sludge, and Biosolids Management: Moving Forward the Sustainable and Welcome Uses of a Global Resource* (Nairobi: UN-HABITAT, 2008). Critiques of biosolids in agriculture include Abby Rockefeller, "Civilization and Sludge: Notes on the History of the Management of Human Excreta," *Capitalism, Nature, Socialism* 9, no. 3 (1998): 3–18; Caroline Snyder's "The Dirty Work of Promoting 'Recycling' of America's Sewage Sludge," *International Journal of Occupational and Environmental Health* 11 (2005): 415–427; Jordan Peccia and Paul Westerhoff, "We Should Expect More out of Our Sewage Sludge," *Environmental Science & Technology* 49 (2015): 8271–8276.

On the historical uses for sludge, see Marta E. Szczygiel, "From Night Soil to Washlet," *ejcjs* 16, no. 3 (2016); Wilson and Flohr's "The Economy of Odure"; *Water 4.0*; Erland Mårald, "Everything Circulates: Agricultural Chemistry and Recycling Theories in the Second Half of the Nineteenth Century," *Environment and History* 8, no. 1 (2002): 65–84; Sarah Jewitt, "Geographies of Shit: Spatial and Temporal Variations in Attitudes towards Human Waste," *Progress in Human Geography* 35, no. 5 (2011): 608–626; Nicholas Goddard, "'A Mine of Wealth'? The Victorians and the Agricultural Value of Sewage," *Geography* 22, no. 3 (1996): 274–290; Schneider's publications include a chapter in *Histories of the Dustheap: Waste, Material Cultures, Social Justice* (Cambridge: MIT Press, 2012) and the book *Hybrid Nature: Sewage Treatment and the Contradictions of the Industrial Ecosystem* (Cambridge: MIT Press, 2011). Resource recovery estimates from Manzoor Qadir et al., "Global and Regional Potential of Wastewater as a Water, Nutrient and Energy Source," *Natural Resources Forum* (2020): 1–12. Weir originally self-published *The Martian* in 2011.

Gone to Pot: For Moule, I referred to *Privies and Water Closets*.

Super Fly: This section draws on and adapts reporting I did in Indonesia and South Africa as part of my European Journalism Centre–supported Waste Not project, particularly for the article "Wie diese Soldaten aus Mist Geld machen können" in *Tagesspiegel*, October 14, 2019 (editing and translation by Sascha Karberg and Richard Friebe), as well as the radio feature "Managing Food Waste with Maggots in Indonesia" for *Deutsche Welle World in Progress*, June 5, 2019, produced by Anke Rasper. (See end for details about the grant.) It also draws on reporting from my earlier *Nature* feature "The New Economy of Excrement," September 13, 2017 (with excellent editing by Kerri Smith), and *Welt-sichten* feature "Der Schatz aus der Toilette," May 2018 (with gratitude to editor Gesine Kauffmann). Thanks also to the Eawag/Waste4Change team in Indonesia for showing me their black soldier fly facility, which processes food waste. For Louisiana story, see G. H. Bradley, "Hermetia illucens L. A Pest of Sanitary Privies in Louisiana," *J Econ Entomol* 23, no. 6 (1930): 1012–1013. Gas-to-protein in J. F. Foster and J. H. Litchfield, "A Continuous Culture Apparatus for the Microbial Utilization of Hydrogen Produced by Electrolysis of Water in Closed-Cycle Space Systems," *Biotechnol Bioeng* 6 (1964): 441 456; John H. Litchfield, "Opportunities and Progress," *Annu Rev Food Sci Technol* 5 (2014): 23–37; Lisa M. Steinberg et al., "Coupling of Anaerobic Waste Treatment to Produce Protein- and Lipid-Rich Bacterial Biomass," *Life Sciences in Space Research* 15 (2017): 32–42; and Silvio Matassa et al., "Can Direct Conversion of Used Nitrogen to New Feed and Protein Help Feed the World?," *Environmental Science & Technology* 49, no. 9 (2015): 5247–5254a.

Shishkov's publication is Shishkov et al., "Black Soldier Fly Larvae Feed by Forming a Fountain around Food," *Journal of the Royal Society, Interface* 16, no. 151, art. no. 20180735 (2019). For background, Martin Hauser and Marshall Woodley, "The Historical Spread of the Black Soldier Fly, Hermetia illucens (L.) (Diptera, Stratiomyidae, Hermetiinae), and Its Establishment in Canada," *Journal of the Kansas Entomological Society* 146 (2015): 51–54; Harinder Makkar et al., "State-of-the-art on Use of Insects as Animal Feed," *Animal Feed Science and Technology* 197 (2014): 1–33; and Helena Čičková et al., "The Use of Fly Larvae for Organic Waste Treatment," *Waste Management* 35 (2015): 68–80. On the troubles faced in Durban, see Maximillian Grau et al., "Viability of a Black Soldier Fly Plant for

Processing Urine Diversion Toilet Faecal Waste," published on *Gates Open Research* (2019). Lalander's publications include Lalander et al., "Faecal Sludge Management with the Larvae of the Black Soldier Fly (Hermetia Illucens)—from a Hygiene Aspect," *Science of the Total Environment* 458–460 (2013): 312–318 and "A Comparison in Product-Value Potential in Four Treatment Strategies for Food Waste and Faeces—Assessing Composting, Fly Larvae Composting and Anaerobic Digestion," *GCB Bioenergy* 10 (2018): 84–91.

Poo Power: For more on Sanivation, see Tyler Karahalios et al., "Human Waste-to-Fuel Briquettes as a Sanitation and Energy Solution for Refugee Camps and Informal Urban Settlements" in *Recovering Bioenergy in Sub-Saharan Africa: Gender Dimensions, Lessons and Challenges*, eds. M. Njenga and R. Mendum (Colombo, Sri Lanka: International Water Management Institute, 2018). On different options for energy recovery, see, for example, Miriam Otoo and Pay Drechsel, eds., *Resource Recovery from Waste: Business Models for Energy, Nutrient and Water Reuse in Low- and Middle-Income Countries* (Oxon, UK: Routledge, 2018), and Ningbo Gao et al., "Thermochemical Conversation of Sewage Sludge: A Critical Review," *Progress in Energy and Combustion Science* 79, no. 100843 (2020). For hydrothermal processing of sewage sludge, P. A. Marrone et al., "Bench-Scale Evaluation of Hydrothermal Processing of Technology for Conversion of Wastewater Solids to Fuels," *Water Environment Research* 90, no. 4 (2018): 329–342; Jacqueline M. Jarvis et al., "Assessment of Hydrotreatment for Hydrothermal Liquefaction Biocrudes from Sewage Sludge, Microalgae, and Pine Feedstocks," *Energy Fuels* 32 (2018): 8483–8493; and Richard L. Skaggs et al., "Waste-to-Energy Biofuel Production Potential for Selected Feedstocks in the Conterminous United States," *Renewable and Sustainable Energy Reviews* 82 (2018): 2540–2651.

Modern Manure: Thanks additionally to David Duest for a tour and explanation of the Deer Island plant. On contaminants in biosolids, there are many references, including the highly cited Chad A. Kinney et al., "Survey of Organic Wastewater Contaminants in Biosolids Destined for Land Application," *Environmental Science & Technology* 40, no. 23 (2006): 7207–7215, and Bradley O. Clarke and Stephen R. Smith, "Review of 'Emerging' Organic Contaminants in Biosolids and Assessment of International Research Priorities for the Agricultural Use of Biosolids," *Environment International* 37, no. 1 (2011): 226–247. The *Boston Globe* report, by David Abel, is titled " 'Forever Chemicals' Are Found in MWRA Fertilizer, Drawing Alarm," December 20, 2019.

Clogged Arteries

Chapter epigraph from Chuck Palahniuk, *Fight Club* (New York: W. W. Norton & Company, 1996).

Potty Break: For a general overview on plastics in wastewater, see Jung Sun et al., "Microplastics in Wastewater Treatment Plants: Detection, Occurrence and Removal," *Water Research* 152 (2019): 21–37; for specifics on microfibers from washing, see Hayley K. McIlwraith et al., "Capturing Microfibers—Marketed Technologies Reduce Microfiber Emissions from Washing Machines," *Marine Pollution Bulletin* 139 (2019): 40–45.

Ultra-Soft Power: "Destroying forests" quote from Katie Notopoulos's "Millennials Are Finally Getting the Giant Roll of Toilet Paper They Deserve" in *BuzzFeed*, June 16, 2019. *The Issue with Tissue* report, authored by Jennifer Skene with contributions from Shelley Vinyard, is from 2019; in June 2020, the NRDC published an update *The Issue with Tissue 2.0*. Colin Beavan wrote about his toilet-paper experience in *No Impact Man: The Adventures of a Guilty Liberal Who Attempts to Save the Planet and the Discoveries He Makes about Himself and Our Way of Life in the Process* (New York: Farrar, Straus and Giroux, 2009).

Garbage In, Garbage Out: The Lusaka study on latrine trash is James Madalitso Tembo et al., "Pit Latrine Faecal Sludge Solid

Waste Quantification and Characterization to Inform the Design of Treatment Facilities in Peri-Urban Areas: A Case Study of Kanyama," *African Journal of Environmental Science and Technology* 13, no. 7 (2019): 260–272. On pit emptying in general, and the Flexcrevator specifically, see David Still, Kitty Foxon, and Mark O'Riordan's three-volume report from 2012, *Tackling the Challenges of Full Pit Latrines*, and Chris Buckley, Rebecca Sindall, and Francis de los Reyes's "Designing Pit Emptying Technologies: Combining Lessons from the Field with Systems Thinking," in *Local Action with International Cooperation to Improve and Sustain Water, Sanitation and Hygiene (WASH) Services: Proceedings of the 40th WEDC International Conference, Loughborough, UK, 24–28 July 2017*, ed. R. J. Shaw, Paper 2826 (2017).

Hot Stuff: I first wrote about sewage-source heat pumps in a *New Scientist* feature called "Smart Heat Nets Fire the Next Energy Revolution," April 15, 2013, conceived of and edited by the amazing Sally Adee. For background on heat recovery, see Silvia Foire and Giuseppe Genon, "Heat Recovery from Municipal Wastewater: Evaluation and Proposals," *Environmental Engineering and Management Journal* 13, no. 7 (2014): 1595–1604; Andrei David et al., "Heat Roadmap Europe: Large-Scale Electric Heat Pumps in District Heating Systems," *Energies* 10 (2017): 578; and Helge Averfalk et al., "Large Heat Pumps in Swedish District Heating Systems," *Renewable and Sustainable Energy Reviews* 79 (2017): 1275–1284.

Junk in the Sewer Trunk: On the history of FOGs, Thomas Wallace et al., "International Evolution of Fat, Oil and Grease (FOG) Waste Management—a Review," *Journal of Environmental Management* 187 (2017): 424–435. On the processes of fatberg formation, see Xia He et al., "Evidence for Fat, Oil, and Grease (FOG) Deposit Formation Mechanisms in Sewer Lines," *Environmental Science & Technology*

45, no. 10 (2011): 4385–4391; Eva Nieuwenhuis et al., "Statistical Modelling of Fat, Oil and Grease (FOG) Deposits in Wastewater Pump Sumps," *Water Research* 135 (2018): 155–167; and Martin A. Gross et al., "Evaluation of Physical and Chemical Properties and Their Interactions in Fat, Oil, and Grease (FOG) Deposits," *Water Research* 123 (2017): 173–182. On gutter oil, see Fangqi Lu and Xuli Wu, "China Food Safety Hits the 'Gutter,'" *Food Control* 41 (2014): 134–138, and Mingming Lu et al., "The Gutter Oil Issue in China," *Proceedings of the Institution of Civil Engineers—Waste and Resource Management* 166, no. 3 (2013): 142–149. Further material from the 2019 and 2020 SwiftComply FOG Summit in Amsterdam.

Giving a Crap

Chapter epigraph from Eleanor Roosevelt's remarks to the United Nations Commission on Human Rights in New York on March 27, 1958, quoted in Suzy Platt, ed., *Respectfully Quoted: A Dictionary of Quotations Requested from the Congressional Research Service* (Washington, D.C.: Library of Congress, 1989). Credit goes to Marni Sommer for applying this quote to the topic of toilets.

A Seat of Privilege: I have previously told the story of this latrine visit in my 2016 *Nature* feature "The Secret History of Ancient Toilets," as well as my June 18, 2014, *Discover* item, "Ancient Rome's Terrorizing Toilets." Plaskow quote in her article "Taking a Break: Toilets, Gender, and Disgust" in *South Atlantic Quarterly* 115, no. 4 (2016). Writer Derek Thompson coined *opulent bathroomification* in a Twitter thread; he also wrote "America Is Overrun with Bathrooms," *The Atlantic,* January 23, 2020. Inequality data from, to name a few, Marielle Snel's "WASH in Prisons Is a Neglected Human Right," IRC *WASH Blog,* June 28, 2020; "WASH in Schools" and "WASH in Health Care Facilities" at Unicef.org; Oxfam's 2016 report *No Relief: Denial of Bathroom Breaks in the Poultry Industry*; and Edward Ongweso, Jr., and Lauren

Kaori Gurley's "Gig Workers Have Nowhere to Pee," *Vice*, January 30, 2020. Toilet scrubbing as women's work in Leah Ruppanner's "We Can Reduce Gender Inequality in Housework—Here's How," *The Conversation*, May 29, 2016; women in the water workforce data in *Women in Water Utilities: Breaking Barriers* (Washington, D.C.: World Bank, 2019). Cattelan quote in "Game of Throne: Maurizio Cattelan's 'America' Comes to the Guggenheim" on the museum's website on September 15, 2016; see also Matthew Bown, "Toilets of Our Time," *TLS*, August 23 and 30, 2019.

Equality. Period.: Of those mentioned in this chapter, Coryton, Weiss-Wolf, Winkler, and many others spoke in the Period Posse webinar series run by the Gender, Adolescent Transitions, and Environment program at Columbia Public Health. Some quotes come from those talks. Bobel's book *The Managed Body* was published in 2019 in New York by Palgrave. For the study in Kenya, Penelope A. Phillips-Howard et al., "Menstrual Needs and Associations with Sexual and Reproductive Risks in Rural Kenyan Females: A Cross-Sectional Behavioral Survey Linked with HIV Prevalence," *Journal of Women's Health* 24, no. 10 (2015): 801–811. For the examples of challenges, the #WASHFail team initiated a Twitter thread on the topic in early 2020. The mentioned artists were featured in Sonam Joshi's article "The Red on Their Canvas? It's Not Paint," *Times of India*, February 23, 2020.

Passing the Smell Test: Much of the detail in this section comes from Starkenmann's as-yet-unpublished manuscript about his experiences, titled *A World Tour of Toilet Smell*. Some quotes from him are from that manuscript. Other sources include Ludwig Brieger's "Ueber die flüchtigen Bestandtheile der menschlichen Excremente," *Journal für Praktische Chemie* 179 (1878): 124–138; Charles Jean-François Chappuis et al., "Quantitative Headspace Analysis of Selected Odorants from Latrines in Africa and India," *Environmental Science & Technology*

49 (2015): 6134–6140; a Wharton School, University of Pennsylvania, case study prepared by Djordjija Petkoski, titled "How Innovation and Partnerships Can Save Lives: Firmenich's Positive Contribution to the Global Sanitation Crisis" (2019); and an Archipel&Co. report called *Malodor & Sanitation Behaviors in Low-Income Settlements* (2019). D'Arbeloff spoke about the business aspects in a 2019 webinar titled "Global Access Strategy and Opportunities in Sanitation."

A Public Domain: For some of the history of public toilets in the United States, see Sophie House, "Pay Toilets Are Illegal in Much of the U.S. They Shouldn't Be," in *CityLab* (now *Bloomberg CityLab*), November 19, 2018. Advocacy group PHLUSH maintains a useful *Public Toilet Advocacy Toolkit*, which includes "A Brief History of Public Toilet Advocacy." "Soulless receptacle" quote from John Metcalfe, "Why Portland's Public Toilets Succeeded Where Others Failed," *CityLab*, January 23, 2012. Kher quote from "How These Pune Sanipreneurs Are Writing a New Script of 'Toilet, Ek Prem Katha,'" by Dipti Nair, in *SMBStory*, April 19, 2019. See also Lezlie Lowe, *No Place to Go: How Public Toilets Fail Our Private Needs* (Toronto: Coach House Books, 2018).

Nuke the Powder Room: Sources for this section include Kogan's chapter "Sex Separation: The Cure-All for Victorian Social Anxiety" and Olga Gershonson's chapter "The Restroom Revolution: Unisex Toilets and Campus Politics," both in *Toilet: Public Restrooms and the Politics of Sharing*. In addition, Brian S. Barnett et al., "The Transgender Bathroom Debate at the Intersection of Politics, Law, Ethics, and Science," *Journal of the American Academy of Psychiatry and the Law* 46 (2018): 232–241; Heath Fogg Davis, "Why the 'Transgender' Bathroom Controversy Should Make Us Rethink Sex-Segregated Public Bathrooms," *Politics, Groups, and Identities* 6, no. 2 (2018); Kathryn H. Anthony and Meghan Dufresne, "Potty Parity in Perspective: Gender and Family Issues in Planning and Designing

Public Restrooms," *Journal of Planning Literature* 21, no. 3 (2007); Ruth Colker, "Public Restrooms: Flipping the Default Rules," *Ohio State Law Journal* 78, no. 1 (2017); and Catherine Joseph and Whitney Odell, "Bathrooms for Humans," *FXCollaborative Podium*, March 20, 2018. The Cuningham Group report on school restrooms is *Inclusive Restroom Design Guide: A Comprehensive Analysis of Inclusive & Gender Specific Restrooms in K–12 Schools* (2020). For more on the Stalled! collaboration's research and ideas, see www.joelsanders architect.com, www.mixdesign.online, and www.stalled.online.

Potty Talk

Quote for epigraph found in *Psychology in the Bathroom*, which references Alan Dundes, "Here I Sit: A Study of American Latrinalia," *Kroeber Anthropological Society Papers* 34 (1966): 91–105.

Cutting the Crap: Thanks to Ari Kamasan, who worked on the World Bank program, for accompanying me in Surabaya, as well as to Risyiana Muthia, for excellent interpretation and other support. Useful references on sanitation in Indonesia, as well as the specific program in which Pak Koen participated, include these World Bank Water and Sanitation Program's publications: Jacqueline Devine, *Introducing SaniFOAM: A Framework to Analyze Sanitation Behaviors to Design Effective Sanitation Programs* (Washington, D.C.: World Bank, 2009); *Results, Impacts, and Learning from Improving Sanitation at Scale in East Java, Indonesia*, Field Note 85200 (Washington, D.C.: World Bank, 2013); and Eduardo Perez et al., *What Does It Take to Scale Up Rural Sanitation?* (Washington, D.C.: World Bank, 2012). Also Nilanjana Mukherjee and Nina Shatifan, *The CLTS Story in Indonesia: Empowering Communities, Transforming Institutions, Furthering Decentralization* (2009); the chapter "Indonesia's Total Sanitation and Sanitation Marketing Program" in Amanda Glassman and Miriam Temin, *Millions Saved: New Cases of Proven Success in Global Health* (Washington, D.C.: Center for Global

Development, 2016); and the Global Delivery Institute's 2015 case study *How to Scale Up Rural Sanitation Service Delivery in Indonesia* by Sarah Glavey and Oliver Haas.

Robert Hume's book is *Thomas Crapper: Lavatory Legend* (Broadstairs, UK: Stone Publishing House, 2009); I drew from his January 6, 2010, *BBC History* article, "The Legend of Thomas Crapper." Eveleigh's book, previously mentioned, is *Privies and Water Closets*.

Poo and Taboo: This section owes a lot to *Psychology in the Bathroom*. For the washlet, I referred to Allen Chun, "Flushing in the Future: The Supermodern Japanese Toilet in a Changing Domestic Culture," *Postcolonial Studies* 5 (2002): 153–170, and the previously mentioned "From Night Soil to Washlet." Sim quotes from TEDxSalford, "Why We Need to Talk about Shit" (2015).

Game of Thrones: The Lusaka trial and related research are published in James B. Tidwell et al., "Effect of a Behaviour Change Intervention on the Quality of Peri-Urban Sanitation in Lusaka, Zambia: A Randomised Controlled Trial," *The Lancet Planetary Health* 3, no. 4 (2019): E187–E196; Tidwell et al., "Theory-Driven Formative Research on On-Site, Shared Sanitation Quality Improvement among Landlords and Tenants in Peri-Urban Lusaka, Zambia," *International Journal of Environmental Health Research* 29, no. 3 (2019): 312–325; and Tidwell et al., "Understanding Demand for Higher Quality Sanitation in Peri-Urban Lusaka, Zambia through Stated and Revealed Preference Analysis," *Social Science & Medicine* 232 (July 2019): 139–147. The block quote is from Tidwell's article "'Talking Shit' Pays Off for Landlords and Tenants in Developing Cities" in *Behavioral Science*, October 22, 2018. Full videos from the project on Ben Tidwell's YouTube channel. General concept of Behaviour Centred Design in Robert Aunger and Val Curtis, "Behaviour Centred Design: Towards an Applied Science of Behaviour Change," *Health Psychology Review*

10, no. 4 (2016): 425–446, and Aunger's book *Reset: An Introduction to Behavior Centered Design* (Oxford: Oxford University Press, 2020).

The Bottom Line: Saito's work at SoAndrewSaito.com.

Life of a Salesman: Lik Telek incident reported in Susan Engel and Anggun Susilo, "Shaming and Sanitation in Indonesia: A Return to Colonial Public Health Practices?," *Development and Change* 45, no. 1 (2014): 157–178.

Epilogue: It Hits the Fan

Slate quote is the title of a May 7, 2020, article by Henry Grabar. Roy's essay is "The Pandemic Is a Portal," *Financial Times,* April 3, 2020.

Funding note: Support for my 2018–2019 Waste Not reporting project, which covered solutions for recovering nutrients from urban organic waste for agriculture for the publications *Deutsche Welle*, *Rethink*, and *Tagesspeigel*, was provided by the European Journalism Centre via its Innovation in Development Reporting grant program, including some travel and reporting costs in Haiti, Indonesia, and South Africa. I have adapted some reporting from that project for this book (specifics appear in the above notes). I also did additional reporting, which appears for the first time in this book, during those trips. Funding for the EJC IDP grants comes from the Bill and Melinda Gates Foundation, which plays no role in selecting the grantees or administering the grants. Some travel to Alaska (including Toolik Field Station) in 2010 and the Boston area in 2017 was covered and organized as part of my participation in the Marine Biological Laboratory Logan Science Journalism Program's Polar Hands-On Laboratory and Biomedical Hands-On Course.

Acknowledgments

I'd like to extend my heartfelt thanks to everyone who signed on to Team Toilet. Agent Alice Martell and editor Ben Lochnen both quickly saw the potential in this project and maintained their enthusiasm throughout. The wonderful folks at Avid Reader Press, particularly Carolyn Kelly, Jessica Chin, Alexandra Primiani, Meredith Vilarello, and Morgan Hoit, remained steadfast even as a pandemic shook their world. People who generously read drafts and gave helpful feedback (plus other invaluable forms of support) at various stages of this project, from proposal to final, include Kim Worsham, Eline Bakker, Jacqueline Ferrand, Polly Shulman, Naomi Lubick, Rhitu Chatterjee, Carolyn Gramling, Jennifer Barr, Risyiana Muthia, Linda Strande, Claire Furlong, Robert Buckley, Maria Lousada Ferreira, Dani Barrington, Alice Miller, and Regan Penaluna. My long-standing science writing "tribe," particularly members Christie Aschwanden and Julie Rehmeyer, provided advice and encouragement. Lexi Pandell and Janet Byrne fact-checked with appropriate ferocity. (Any errors in this book, however, are entirely my own responsibility.)

In addition, a wide variety of friends and family kept me grounded, sent me useful and amusing links and photos, suggested contacts, and let me jabber on about toilets in public places. Childcare providers Christine van Peeren and Kim Buters helped me make enough space in my life for this work. My parents, Penny and Mitch Wald, showed up when and where I asked them to. My partner, Cyril Emery, did a lot of everything, from library reference to midnight reassurances to changing bazillions of diapers on long-haul flights to reporting destinations. My wonderful kid, Ephraim, was in many ways born alongside this book and, like this book, has taught me a lot about myself and priorities and poop.

I also owe a debt to the many journalists and experts whose work on toilets and toilet waste has informed my own, and sources who have let me and other curious investigators into their bathrooms and answered sometimes uncomfortable questions. They all made this book better, though, to my regret, I could not manage to make these pages accommodate nearly all of their stories and names.

Image Credits

Index

Page numbers in *italics* refer to images.

About the Author

Chelsea Wald has repeatedly plunged into the topic of toilets since 2013, when editors first approached her to write about the latent potential in our stagnating sanitation infrastructure. She has won several awards and reporting grants, including from the Society of Environmental Journalists, the European Geosciences Union, and the European Journalism Centre. She lives with her family in the Netherlands, in a region renowned for its water-related innovations.